A CULTURAL HISTORY OF CHEMISTRY

VOLUME 5

A Cultural History of Chemistry
General Editors: Peter J.T. Morris and Alan J. Rocke

Volume 1
A Cultural History of Chemistry in Antiquity
Edited by Marco Beretta

Volume 2
A Cultural History of Chemistry in the Middle Ages
Edited by Charles Burnett and Sébastien Moureau

Volume 3
A Cultural History of Chemistry in the Early Modern Age
Edited by Bruce T. Moran

Volume 4
A Cultural History of Chemistry in the Eighteenth Century
Edited by Matthew Daniel Eddy and Ursula Klein

Volume 5
A Cultural History of Chemistry in the Nineteenth Century
Edited by Peter J. Ramberg

Volume 6
A Cultural History of Chemistry in the Modern Age
Edited by Peter J.T. Morris

A CULTURAL HISTORY OF CHEMISTRY

IN THE NINETEENTH CENTURY

VOLUME 5

Edited by Peter J. Ramberg

BLOOMSBURY ACADEMIC
LONDON • NEW YORK • OXFORD • NEW DELHI • SYDNEY

BLOOMSBURY ACADEMIC
Bloomsbury Publishing Plc
50 Bedford Square, London, WC1B 3DP, UK
1385 Broadway, New York, NY 10018, USA
29 Earlsfort Terrace, Dublin 2, Ireland

BLOOMSBURY, BLOOMSBURY ACADEMIC and the Diana logo are trademarks
of Bloomsbury Publishing Plc

First published in Great Britain 2021
Paperback edition published in 2025

Copyright © Bloomsbury Publishing Plc, 2025

Cover design: Rebecca Heselton
Cover image © Universal History Archive/UIG/Bridgeman Images

All rights reserved. No part of this publication may be reproduced or transmitted
in any form or by any means, electronic or mechanical, including photocopying,
recording, or any information storage or retrieval system, without prior permission
in writing from the publishers.

Bloomsbury Publishing Plc does not have any control over, or responsibility for,
any third-party websites referred to or in this book. All internet addresses given
in this book were correct at the time of going to press. The author and publisher
regret any inconvenience caused if addresses have changed or sites have ceased
to exist, but can accept no responsibility for any such changes.

A catalogue record for this book is available from the British Library.

A catalog record for this book is available from the Library of Congress.

ISBN: PB: 978-1-3505-5217-3
Pack: 978-1-3505-5229-6
ePUB: 978-1-3502-5155-7
ePDF: 978-1-3502-5154-0

Series: The Cultural Histories Series

Typeset by Integra Software Services Pvt. Ltd.
Printed and bound in Great Britain

To find out more about our authors and books visit www.bloomsbury.com
and sign up for our newsletters.

CONTENTS

LIST OF ILLUSTRATIONS — vii
LIST OF TABLES — xi
SERIES PREFACE — xii

Introduction: Creating Modern Chemistry — 1
Peter J. Ramberg

1 Theory and Concepts: Atomism, Structure, and Affinity — 41
   Trevor Levere

2 Practice and Experiment: Analysis, Synthesis, and Paper Tools — 67
   Yoshiyuki Kikuchi

3 Laboratories and Technology: Continuity and Ingenuity in the Workplace — 91
   Amy A. Fisher

4 Culture and Science: Chemistry Spreads Its Influence — 117
   Agustí Nieto-Galan and Peter J. Ramberg

5 Society and Environment: Increased Access for Women, Growing Consumerism, and Emerging Regulation — 139
   Peter Reed

6 Trade and Industry: New Demands, New Processes, and the Emergence of Science-Based Chemical Industry — 167
   Anthony S. Travis

7 Learning and Institutions: Emergence of Laboratory-Based Learning,
   Research Schools, and Professionalization   191
   Peter Reed

8 Art and Representation: The Rise of the "Mad Scientist"   217
   Joachim Schummer

NOTES   239
BIBLIOGRAPHY   241
LIST OF CONTRIBUTORS   260
INDEX   261

LIST OF ILLUSTRATIONS

0.1 The type theory. From left to right, the water, ammonia, hydrogen, and hydrochloric acid types. Drawing by Peter J. Ramberg   9

0.2 Crum Brown's notation indicating the difference between two isomers of propyl alcohol. Drawing by Peter J. Ramberg   11

0.3 Charles Friedel (seated, sixth from the left), Cannizzaro (seated, fifth from left), Baeyer (seated, fourth from left) at the 1892 Geneva Congress for Reform of Chemical Nomenclature. Photograph, 1892. Courtesy of the Wellcome Collection   23

0.4 An 1861 cartoon that comments on the spectacular success of the new synthetic dyes (*Punch*, November 23, 1861: 212). Courtesy of the Sidney M. Edelstein Center, The Hebrew University of Jerusalem, Israel   29

0.5 Rutherford's illustration of transmution of elements by α-, β-, and γ- radiation (Rutherford 1906). Courtesy of the Oesper Collection, University of Cincinatti   31

0.6 Advertisement for radium-containing medicinal products, 1916. Courtesy of the Wellcome Collection   35

1.1 Wollaston's slide rule of chemical equivalents (Wollaston 1814). Courtesy of the Science History Institute, Philadelphia   45

1.2 Kekulé's "sausage" formulas. Bonds between atoms are shown by those points of atoms which touch vertically (Kekulé 1861: 162). Courtesy of the Oesper Collection, University of Cincinatti   54

1.3 Formulas for ethane, ethylene, and acetylene using Crum Brown's symbolism. Drawing by Peter J. Ramberg — 54

1.4 Kekulé's early cyclic formulas for benzene (Kekulé 1867: 4, 22, 6–7). Courtesy of the Science History Institute, Philadelphia — 54

1.5 Van 't Hoff's depiction of the tetrahedral carbon atom. The top is a mirror image of the bottom, and the two tetrahedra are not superimposable (van 't Hoff 1874). Courtesy of the Oesper Collection, University of Cincinatti — 56

1.6 Mendeleev's first Periodic Table of Elements, 1969. Photograph by Ann Ronan Pictures/Print Collector/Getty Images — 59

2.1 The blowpipe in use (Mawe 1825: 4). Courtesy of the Wellcome Collection — 69

2.2 Berzelius' laboratory equipment (Berzelius 1841). Courtesy of the Bayerische Staatsbibliothek München, Chem. 30 d-10,a, um:nbn:de:bvb:12-bsb10072019-2 — 71

2.3 Gay-Lussac and Thenard's volumetric apparatus for organic analysis (Gay-Lussac and Thenard 1811). Courtesy of the Osaka Prefectural Central Library, Japan — 73

2.4 Berzelius' gravimetric apparatus for organic analysis (Berzelius 1814: 404). Courtesy of the Natural History Museum, London — 74

2.5 Hofmann's "glyptic" (sculpted) models (Hofmann 1865: 426). Courtesy of the Natural History Museum, London — 79

2.6 Kekulé's tetrahedral models of six carbon atoms forming a benzene ring. Courtesy of the University of Ghent — 80

3.1 An engraving by Alfred William Warren (1822) depicting a laboratory designed mostly for solitary work. Courtesy of the Science History Institute, Philadelphia — 94

3.2 Lithograth of Justus von Liebig's analytical laboratory in Giessen, *ca*. 1845. Contrast with the laboratory of Figures 3.1 and 7.1. Courtesy of the Sidney M. Edelstein Library, The National Library of Israel — 98

3.3 An illustration of assorted furnaces, including fixed and portable furnaces (Accum 1824). Courtesy of the Science History Institute, Philadelphia — 105

LIST OF ILLUSTRATIONS ix

3.4 An etching by Walter Ouler (1875) depicting a chemist watching over a distillation process. Courtesy of the Science History Institute, Philadelphia — 106

3.5 The spectroscope invented by Bunsen and Kirchhoff. One tube allowed light from the combustion reaction to enter the prism apparatus. A telescope provided the experimenter with a view of the spectrum. A second telescope provided the experimenter with a view of a scale with which to measure the relative location of each line in the spectrum. Photograph by Photo 12/Universal Images Group via Getty Images — 108

3.6 Liebig's *Kaliapparat*. On the left is a small tray to hold burning coal that would heat the sample in an enclosed glass tube. The water produced during combustion was absorbed by calcium chloride in the tube immediately to the right of the tray. The heart of the apparatus was the five-bulb fixture on the right that captured carbon dioxide with a saturated solution of aqueous potassium hydroxide. Courtesy of the Wellcome Collection — 111

3.7 The 1913 Franz Schmidt & Haensch polarimeter, a modified version of the Jellett–Cornu polarimeter. Courtesy of the Science History Institute, Philadelphia — 113

4.1 Henri Sainte-Claire Deville demonstrating an experiment in the inorganic chemistry laboratory at the *École Normale Supérieure*, Paris, 1878. Courtesy of the Wellcome Collection — 134

4.2 Michael Faraday lecturing at the Royal Institution with an audience of many women and some British royals. Photograph by Time Life Pictures/Mansell/The LIFE Picture Collection via Getty Images — 135

5.1 The opening of the Great Exhibition on May 1, 1851. Engraved by H. Bibby, 1851. Courtesy of the Wellcome Collection — 141

5.2 Marie Skłodowska Curie in her laboratory. Photograph by Time Life Pictures/Mansell/The LIFE Picture Collection via Getty Images — 145

5.3 Salvarsan treatment kit for syphilis, Germany, 1909–1912. Photograph by Science and Society Picture Library/Getty Images — 149

5.4 Carbolic steam spray made by D. Marr and used by Joseph Lister (1827–1912). Photograph by Science and Society Picture Library/Getty Images — 155

5.5	Acid tower at Messrs. John Hutchinson & Co., Widnes, in 1860 (Lunge 1886). Courtesy of Peter Reed	159
6.1	The cluster of alkali and allied chemical factories situated in Widnes, Cheshire, northwest England, close to the River Mersey and canals, which facilitated transport and in many cases acted as waste sinks. The several tall chimneys dispersed corrosive gases into the surrounding atmosphere. Probably late nineteenth century. Courtesy of the Sidney M. Edelstein Center, The National Library of Israel	171
6.2	Manufacture of coal gas by destructive distillation of coal. From mid-century the waste tar from the coal gas-making processes was transformed into synthetic dyestuffs, and the waste ammonia into ammonium sulfate fertilizer (Accum 1819). Courtesy of the Sidney M. Edelstein Library, The National Library of Israel	173
6.3	The Ludwigshafen factory of BASF (Badische Anilin- & Soda-Fabrik), located on the west bank of the River Rhine, late nineteenth century. In addition to synthetic dyes, the factory manufactured its own alkali and acids. Courtesy of the Sidney M. Edelstein Library, The Hebrew University of Jerusalem, Israel	175
6.4	Shoveling dry Chilean saltpeter (sodium nitrate), from evaporation pans, Arica, Chile. This photograph was taken in 1925, but the scene is typical of the large-scale nitrate industry in the late nineteenth century. Photograph by Bettmann/Getty Images	182
7.1	Faraday's laboratory at the Royal Institution. Photograph by Picture Post/Hulton Archive/Getty Images	193
7.2	The qualitative analysis laboratory, Owens College, Manchester, 1873 (Roscoe 1906). Courtesy of Peter J.T. Morris	199
7.3	Ira Remsen (center) and students in a chemistry laboratory, Johns Hopkins University, 1890. Photograph by JHU Sheridan Libraries/Gado/Getty Images	204

LIST OF TABLES

5.1	Common Adulterants from Hassall's Evidence to 1855 Select Committee	152
5.2	Population Increase for Manchester and Liverpool	162
5.3	Main Industries and Their Pollutants	163

SERIES PREFACE

A Cultural History of Chemistry examines the history of chemistry and its wider contexts from antiquity to the present. The series consists of six chronologically defined volumes, each volume comprising nine essays; these fifty-four contributions were written and/or edited by a total of fifty scholars, of ten different nationalities. Of Bloomsbury's many six-volume *Cultural Histories* currently in print, this is the first in the physical or natural sciences; it is also the first multivolume history of chemistry to appear since James Riddick Partington's four-volume *History of Chemistry*, concluded more than fifty years ago. It is distinguished, among other qualities, by its endeavor to take the subject from antiquity right to the present day.

This is not a conventional history of chemistry, but a first attempt at creating a cultural history of the science. All cultures, including the various branches of natural science, consist of mixed constructs of social, intellectual, and material elements; however, the cultural-historical study of chemistry is still in an early stage of development. We hope that the accounts presented in these volumes will prove useful for students and scholars interested in the subject, and a starting point for those who are striving to create a more fully developed cultural history of chemistry.

Each volume has the same structure: starting with an interpretive overview by the volume editor(s), the eight succeeding chapters explore for each respective era in chemistry its theory and concepts; practice and experiment; laboratories and technology; culture and science; society and environment; trade and industry; learning and institutions; and art and representation. Readers therefore have the option to read multiple chapters in a single volume, thus learning about the cultures of chemistry in a single era; or they may prefer instead to read corresponding chapters across multiple volumes, learning about (e.g.) the art and representations of chemistry through the ages. Though the scope is global, major emphasis is placed on the Western tradition of science and its contexts.

Whether read synchronically or diachronically, in any multiauthor undertaking like this one readers will inevitably notice overlaps and repetitions, conflicting historical interpretations, and (despite the magnitude of the project) occasional gaps in coverage. These are inescapable consequences, but they actually offer advantages to the reader, both in making each chapter closer to self-contained, and in demonstrating the dynamism of the discipline; like science itself, the study of its history is ever contested and incomplete.

Chemistry has been called the "central science," due to its fundamental importance to all the other physical and natural sciences. It is the archetypical science of materials and material productivity, and as such it has always been deeply embedded in human industry, society, arts, and culture, as these volumes richly attest. The editors and authors hope that *A Cultural History of Chemistry* will be of great interest and enjoyment not just to chemists and specialist historians of science, but also to social, economic, intellectual, and cultural historians, as well as to other interested readers.

Peter J.T. Morris and Alan J. Rocke
London (UK) and Cleveland (USA)

Introduction: *Creating Modern Chemistry*

PETER J. RAMBERG

On March 11, 1890, chemists at a special meeting of the German Chemical Society gathered for an elaborate banquet in Berlin. The guest of honor was August Kekulé, who twenty-five years earlier had suggested that the chemical structure for benzene was a ring of six carbon atoms, a theory that had proven to be an extraordinarily productive tool for organic chemists. Guests were treated to a nine-course meal, numerous toasts, and congratulatory speeches that lauded the success of Kekulé's theory. In one of the toasts, August Wilhelm von Hofmann proclaimed that in the time since Kekulé proposed the ring structure, organic chemistry had metaphorically grown from a small, shallow creek, into a "deep and mighty river, across which one could scarcely even see ... " and Kekulé's benzene formula provided the "soil from which, to our astonishment, we saw the organic chemistry with which we were familiar grow tall anew" (Hofmann in Rocke 2010). In his own speech, Kekulé reminisced that he had initially conceived of the benzene theory, and of his original idea that led to the idea of chemical structure, in two bursts of inspiration that had come to him in dream-like states.

While Hofmann and the other speakers at the "Benzolfest" (benzene celebration) might have been understandably prone to exaggeration and self-congratulation, they knew first-hand that organic chemistry had seen spectacular growth since the 1860s due to the success of the structural theory of organic chemistry. Kekulé had first elaborated the outlines of structural theory in the late 1850s, and had then used it to develop his benzene theory. By the 1860s, chemists quickly realized the strength of the structural theory for bringing order to the thousands of known organic compounds and predicting still more compounds. In the world of nineteenth-century chemical theory, the late 1850s would be the crucial turning point, or a "quiet revolution," when most chemists quickly united behind a single vision of the molecular world (Rocke 1993).

The emergence of the structural theory of chemistry during the 1860s, like the appearance of evolution in biology and electromagnetic theory in physics, was one of the great intellectual accomplishments of nineteenth-century science, an essential component in the explosive growth of chemistry during the nineteenth century as a science, as a profession, and as an industry. A chemist from 1815 transported to 1914 might recognize the general outlines of chemical theory and practice in 1914, but they would likely be astounded by the changes. The number of practicing chemists had dramatically increased, and the vast majority of these chemists were now trained professionally *as chemists*, publishing articles in journals devoted exclusively to chemistry, or to a subdiscipline of chemistry. The number of new chemical compounds had increased exponentially, and the commercial wealth generated by the chemical industry was equally staggering. Universities in all industrialized countries had large buildings devoted to chemical teaching and research, and large chemical firms had laboratories dedicated to chemical research and development.

There are numerous reasons why chemistry grew and changed so dramatically during the nineteenth century. By mid-century, chemists had developed a new institution – the teaching–research laboratory – for the "mass production" of trained chemists that would become a template for training in all the sciences. The increasing number of chemists met a growing need for the increased demands of the marketplace – the need for steel, copper, dyes, oil, soap, pharmaceuticals, and other material goods, all of which required trained chemists. Chemists also developed increasingly powerful theoretical and practical tools based on the modern atomic theory introduced by John Dalton between 1803 and 1810. The concept of rational formulas and the development of a unique visual language to express those formulas gave chemists increasing, although by no means complete, control over nature. The new structure theory would make possible the new artificial dye industry that would in turn influence fashion and consumer desires.

The theoretical culture of chemistry would become increasingly allied to physics by the end of the century, when chemists and physicists struggled to understand radioactivity and other novel phenomena. By 1914, simple Daltonian atomism had disappeared, replaced by atoms with subatomic structures composed mostly of empty space and modified with the new concept of isotopes. Atoms of the new radioactive elements, furthermore, were no longer stable, enduring entities, but capable of transforming themselves into atoms of different elements. The phenomenon of radioactivity would quickly move far beyond chemistry and physics, influencing biology and geology, and it would take hold of the public imagination.

CONSOLIDATION OF ATOMISM

By the early nineteenth century, chemists had developed three primary and interrelated goals: first, to determine the elemental composition of substances using Antoine-Laurent Lavoisier's system from the late eighteenth century; second, to use Dalton's atomic theory to establish atomic weights for the known elements and determine the proportion by weight of each element in pure substances; third, to express those proportions in formulas of some kind. The Swedish chemist Jacob Berzelius, one of the early champions of Daltonian atomism, set the theoretical framework for reaching each of these goals. The discovery of current electricity by Alessandro Volta in 1800 and its early application to chemistry suggested to Berzelius and others that chemical affinity must be electrical in nature. When exposed to electrical current, the elements that made up compounds would collect at either the positive or negative pole of a battery, and Berzelius suggested that each element was either "electronegative" or "electropositive." Compounds would therefore consist of a positively charged piece and a negatively charged piece, held together by the forces of electrostatic attraction, a theory that came to be known as "electrochemical dualism." This composition could then be expressed in chemical formulas, using a new notation that Berzelius introduced in 1814, still used in slightly modified form today, that represents atoms with letters and the proportions of atoms by superscript (now subscript) numbers.

By 1815, Daltonian atomism had already made significant headway among European chemists as a method for ascertaining chemical formulas. There was no doubt that Dalton's main idea, that elements combined together in fixed weight proportions, was true, but there were two ways of interpreting Dalton's theory. The first interpretation, physical atomism, held by Dalton himself, Thomas Thomson (Dalton's earliest supporter in Britain), and Berzelius, assumed that these proportions were caused by the weights of the different atoms themselves – solid spheres that had the same relative weights as the macroscopic weight ratios. Under the second interpretation, chemical atomism, chemists assumed that Dalton's weight ratios were correct and useful, but made no claims about the physical nature of the atoms themselves. While chemists remained divided about physical atomism throughout the century, all chemists adhered to various versions of chemical atomism (Rocke 1984). Chemical atomism was also widely adopted in France, and German chemists, influenced by the Romantic tradition that eschewed materialism and supported skepticism about atomism, were nonetheless enthusiastic about the value of Dalton's theory as a formalized set of stoichiometric rules.

However, many chemists rejected physical atoms on methodological and metaphysical grounds, because they found no compelling reason to assume that the cause of the combining proportions was the weight of the atoms themselves. William Hyde Wollaston subsequently called Dalton's atoms "equivalents," a term that referred only to empirically measurable weight proportions. Many nineteenth-century chemists adopted the numerical values of Wollaston's "equivalent weights" for the various chemical elements (e.g. H = 1, C = 6, O = 8, N = 14, etc.) and employed them to derive molecular formulas in the same way as the advocates of relative "atomic weights" (e.g. H = 1, C = 12, O = 16, N = 14, etc.). For this reason they are distinct from the way the word "equivalent" is used by present-day chemists.

Humphry Davy was also a critic of physical atomism, and wrote that Dalton was "too much of an *Atomic Philosopher* [and] has often indulged in vain speculation; and the essential and truly useful part of his doctrine, the expression of the quantities in which bodies combine, is perfectly independent of any views respecting the ultimate nature either of matter or its elements" (Davy 1840; Rocke 1984; Davy 1968; Rocke 2018a). Davy was convinced that the actual number of elements must be much smaller than the long list begun by Lavoisier and used by Dalton, and he felt in his heart that there should be only perhaps two or three elements. William Prout went further, suggesting that because Dalton's atomic weights were dominated almost completely by whole numbers, hydrogen (the lightest element) must be the only real element, or "protyle," each of Dalton's elemental atomic weights being an integral multiple of the hydrogen atom. "Prout's hypothesis," as it came to be known, would hover in the back of chemists' minds throughout the nineteenth century, although more accurate atomic weight measurements indicated that some elements had weights almost directly between two integers (the weight of chlorine, for example, is 35.45; Brock 1985). Yet Davy and other skeptics of physical atomism saw great value in Dalton's method for calculating the weight proportions of elements in compounds. According to Davy, Dalton's laws resembled Kepler's laws of planetary motion that awaited their Newton to establish their true cause.

By 1820, chemists throughout Europe had come to embrace chemical atomism as a tool for constructing chemical formulas, but competing systems of equivalent and atomic weights led to different formulas. Constructing a single set of atomic weights upon which all chemists could agree remained problematic. In order to calculate his atomic weights, Dalton had assumed that atoms would combine in the simplest manner possible. Joseph-Louis Gay-Lussac assumed that equal volumes of gases contained an equal number of atoms in order to calculate molecular weights, an assumption that led to a different set of weights. Adding to the confusion, chemists did not agree on how to set the initial weight to set the scale (for example, hydrogen as 1 or oxygen as 100). The many systems for calculating atomic weights were often, but not always, divided along national lines. By mid-century, the result was many different (but internally consistent) sets of atomic weights across Europe; this led to many different possible formulas for the same compound depending on which version of atomic weights was being assumed. In the first part of his textbook published in 1859, for example, Kekulé listed sixteen different formulas for acetic acid.

The key for resolving the problem had already been suggested in 1811 by Amedeo Avogadro. The competing systems of atomic weights could be collapsed into a single system if chemists assumed that individual particles of the most common gases – hydrogen, oxygen, and nitrogen – consisted of two atoms. At the time, Avogadro's hypothesis was ignored, primarily because it was difficult to understand how two atoms with the same electronegativity (both atoms of hydrogen would be positive, for example) could bond with each other. Avogadro's hypothesis would be revived by Stanislao Cannizzaro at the Karlsruhe Congress of 1860, one of the first international conferences in the sciences. Kekulé and Adolphe Wurtz had convened the Congress for the purpose of resolving the issue of atomic weights. Although the immediate results of the meeting disappointed its organizers, Cannizzaro had distributed a pamphlet there, based on his own attempts to organize and present atomic weights to his students, that explained how Avogadro's hypothesis would lead to a consistent set of atomic weights, and most chemists were soon united behind a single set of atomic weights (Nye 1984).

The final agreement on atomic weights would eventually lead to one of the most significant accomplishments in the unification of chemical phenomena – the periodic law and the periodic table to visualize the law. Early in the nineteenth century, chemists had already recognized the repetition of certain chemical properties among groups of elements. Between 1817 and 1819, Johann Wolfgang Döbereiner noticed that certain elements could be grouped as "triads," for example, lithium, sodium, and potassium had similar properties and sodium's atomic weight was the average of lithium and potassium. What would later become the periodic law was also noted in different ways in the 1860s: John Newlands' "law of octaves" (1863), Gustav Hinrichs' numerological speculation about the recurring sizes and weights of the atoms (1863), and Lothar Meyer's 1864 table that correlated a repetition of valence with atomic weight (Zapffe 1969; Brock 1993).

The most lasting statement of the periodic law would come from Dmitrii Mendeleev in 1869 as the result of writing one of the first original Russian language multivolume general chemistry textbooks. Motivated by contractual reasons to finish the second volume of the text as quickly as possible, and seeking an effective pedagogical tool, Mendeleev settled on using atomic weights as the one true defining constant for each element (Gordin 2004: 24–7). When he ordered the elements according to their weight, the regularities in valence and properties noted by earlier chemists became clear and he arranged all of the elements into a table to express this periodic law. Mendeleev's insistence on the primacy of atomic weights would prove advantageous, as it allowed him, for the first time, to make bold predictions about not only the existence of new elements, but also their atomic weights, densities, and valences. At the same time, his insistence on the validity of the periodic law compelled him to insist that a few atomic weights had been incorrectly measured.

The eventual success and widespread adoption of the periodic table resulted from the isolation of the predicted elements and the discovery that their properties were close to Mendeleev's predictions (Bensaude-Vincent 1986; Brush 1996). Although Mendeleev's table would prove successful, it briefly met some difficulties in the 1890s, when William Ramsay and John William Strutt (Lord Rayleigh) announced the isolation of chemically inert gases that did not fit within the existing table. In 1898, Ramsay argued convincingly using the periodic law that the new gases belonged in a new column of the periodic table (Hirsh 1981; Brock 1993).

Despite the success of the atomic theory in bringing order to chemical phenomena, and despite the final agreement on atomic weights, there remained skepticism and debate about the existence of physical atoms up to the end of the century. In Britain, Benjamin Brodie was extremely skeptical of the speculative, mechanical picture of molecules given by drawings of molecules on paper or by models made of "balls and wires." Edward Frankland and Alexander Williamson were reluctant to commit to physical atoms, but vigorously defended the utility of molecular models or paper drawings as heuristic devices. The extraordinary success of the atomic theory since Dalton, argued Williamson, led "inexorably to the conclusion that matter consists of chemical atoms ... " but said nothing about "whether our elementary atoms are in the their nature indivisible, or whether they are built up of smaller particles" (Williamson in Rocke 1984: 318).

In France, Adolphe Wurtz was nearly alone in strongly defending the atomic theory against the many influential French skeptics of physical atoms. These skeptics were led by Marcellin Berthelot, who exerted a strong influence on French science, particularly after Wurtz's death in 1884, in hindering atomistic ideas in French higher education. Berthelot wanted chemistry to be independent of physics and free of unnecessary hypotheses, including physical atomism. Berthelot's goal was also aided by the prominent views in France of Henri Poincaré and Pierre Duhem, who advocated a conventionalism in which the actual existence of physical atoms was deemed unnecessary for the success of science (Rocke 2000a).

In Germany, the atomic debates came to a head at the 1895 meeting of the German Society of Naturalists and Physicians in Lübeck. At the meeting, the physicist Georg Helm and the chemist Wilhelm Ostwald both advocated an anti-atomistic "energeticist" worldview, in which matter was "imaginary, an entity that we have rather imperfectly constructed to represent that which persists through the flux of appearances" (Ostwald in Rocke 1984: 328). The organizer of the meeting, Johannes Wislicenus, had also arranged for lectures to defend atomism, and both physicists and chemists at the meeting argued strongly against Helm and Ostwald's energeticism, invoking both the kinetic theory of gases, and the well-established existence of stoichiometric laws (Deltete 1999). The latter were clearly the weakness in the energeticists' viewpoint, and like Berthelot, even Ostwald had accepted chemical atomism. The Lübeck conference failed to reach any kind of consensus, but within the next ten years, the discovery of x-rays, the electron, subatomic structure, and finally Albert Einstein's 1905 paper explaining Brownian motion would eventually convince everyone, even Ostwald, of the viability of physical as well as chemical atomism (Rocke 1984).

THE RISE OF ORGANIC CHEMISTRY

"Organic chemistry," contrasted with "mineral chemistry," originally referred to the chemistry of substances derived more or less directly from plants and animals, including the roles of those substances in plant and animal physiology and anatomy. This original conception of organic chemistry was transformed radically during the 1830s into an entirely different science; interest in the origin of organic compounds was deemphasized in favor of studying their reactivity as chemicals (Klein 2003). As the chemistry of these compounds became more understood, the number of artificial compounds increased exponentially. In 1800, chemists knew only a few dozen organic compounds, but around 1820, this number began to climb steadily, doubling roughly every thirteen years (Schummer 1997). Already by the 1830s, the artificial "organic" compounds outnumbered the natural ones, and by the 1840s, chemists explicitly redefined organic chemistry to mean the chemistry of all compounds containing carbon, whether natural or artificial. Further complicating matters, all of these compounds consisted of only a few elements – mostly carbon, hydrogen, oxygen, and nitrogen.

Between 1815 and 1830, Berzelius organized the basic theoretical framework for bringing order to the growing number of organic compounds. First, Berzelius was convinced of the unity of chemical laws, that organic and inorganic compounds alike should follow Dalton's laws of definite proportions. The proportions in organic compounds were more complex than those in inorganic compounds, but that complexity could be understood by assuming that formulas for organic compounds could be found by drawing analogies to the better-known inorganic compounds. Second, in 1830, Berzelius introduced the concept of "isomerism" to explain the numerous compounds that had identical elemental compositions but different chemical properties. When even more such pairs of compounds began to be discovered, such as silver fulminate and silver cyanate, urea and ammonium cyanate, and tartaric and racemic acid, it became clear that composition by itself could not explain chemical properties (Rocke 1984). Third, Berzelius suggested that the different properties of isomers could be explained by a different "arrangement" of atoms in the molecule. This arrangement would then be depicted by "rational" formulas that would group atoms – facilitated greatly by using his own notation – in different ways to emphasize different parts of the molecule. For example (using modern formulas), water could be written H_2O, or HHO or HOH, or ethanol (C_2H_6O) could be written either C_2H_5OH or $C_2H_4 \cdot H_2O$. Each rational formula emphasized a different hypothesis regarding the groupings of the atoms within the molecule. Berzelius' framework for understanding organic compounds proved to be remarkably enduring, as the search for "arrangement" and "rational formulas" to explain isomerism drove much of chemical theory throughout the entire nineteenth century.

Although chemists were in broad agreement that they should express a compound's properties by rational formulas that depicted the arrangement of atoms, they did not all agree about what the best "rational" formula might be. One approach to rational formulas, the radical theory, emerged primarily in Germany. In 1832, Justus Liebig and Friedrich Wöhler published a landmark paper on the derivatives of the oil of bitter almonds (now called benzaldehyde) in which they concluded that each derivative contained the benzoyl "radical," a group of atoms with the same elemental composition (Liebig and Wöhler in Benfey 1963). The presence of such radicals suggested that organic and inorganic compounds followed the same principles of electrochemical dualism, with two components held together by electrical forces. Organic molecules contained complex radicals, while the radicals in inorganic molecules were simple; that was the only difference.

The main alternative to the radical theory would come from France, and was inspired by an 1827 incident at a soiree held at the Tuileries palace, where (it was said) the guests of King Charles X complained about acrid fumes created by the candles in the chandeliers. The young chemist Jean-Baptiste Dumas was asked to analyze the candles, and he found that the fumes were hydrochloric acid generated from chlorine in the wax that had been bleached. This chance encounter led Dumas into an extensive study of chlorination of organic compounds with the startling conclusion that chlorine, an electronegative element, could replace hydrogen, an electropositive element, without changing the fundamental chemical properties of the compound. Dumas found the same result in the much simpler compound acetic acid, where the hydrogen could be replaced by chlorine without removing its acidic properties.

Dumas concluded that something fundamental in the acetic acid molecule must convey the property of acidity, and proposed that both acetic acid and chloroacetic acid belonged to the same chemical "type." Classification by type involved identifying the primary chemical property of a compound (whether it was acidic or basic, for example), and rational formulas were then constructed around this central property. In the 1840s, Auguste Laurent suggested that each type had a fundamental nucleus of carbon atoms that produced those properties that allowed hydrogens to be substituted without changing the fundamental property of the compound. By the 1850s, there were several types based on analogy with water, ammonia, hydrogen, and hydrochloric acid (Figure 0.1).

The implications of the type theory were unacceptable to the proponents of the radical theory. Under the type theory, if a compound had multiple chemical properties, it could be represented by multiple different rational formulas, depending on which chemical property was represented. However, defenders of the radical theory insisted that each compound should have only a single formula, and regarded the putative relativism of the type theory as unacceptable.

At the Benzolfest, Kekulé recounted the emergence of these two differing approaches to arrangement by comparing them to two branches of the same river that had diverged in the 1830s. By the 1850s, the travelers on each branch had begun gradually approaching each other but were still divided by "a thick underbrush of misunderstandings." Kekulé himself had traveled in both branches of this river, and was one of the first to realize that the two competing approaches to organic chemistry could be combined to produce a more productive theory (Rocke 2010: 304). Kekulé's achievement was to combine the alternative visions of rational formulas by insisting that a compound could have only one rational formula, but that formula should be capable of expressing multiple chemical properties.

Building on the new concept of valence (that atoms of each element could combine only with a specified number of atoms), Kekulé suggested that carbon

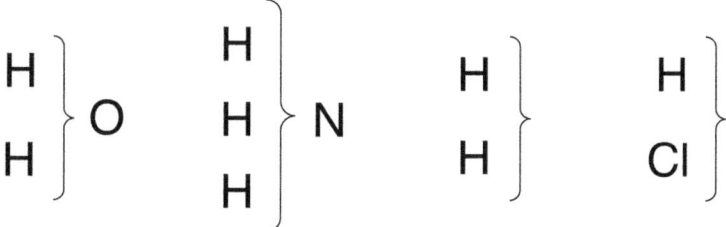

FIGURE 0.1 The type theory. From left to right, the water, ammonia, hydrogen, and hydrochloric acid types. Drawing by Peter J. Ramberg.

could combine with four atoms, and argued that valence was an invariant number for each element. In the molecule of every organic compound every atom must have all of its valences satisfied. Carbon could also satisfy its valence with other carbon atoms, and carbon atoms could form multiple bonds (a consequence of the invariance of carbon atom's valence). The result was the most comprehensive means of reaching Berzelius' original goal – explaining isomers by the arrangement of atoms in the molecule, where arrangement was how the atoms were connected following the rules of valence. There could be only one rational formula for each compound, as the radical theorists demanded, but radicals were now defined simply as the unchanged portion of the molecule in any given reaction. Multiple chemical properties could be represented simultaneously in the same formula.

Kekulé's theory would prove extraordinarily powerful. It explained existing isomers and enabled prediction of new compounds simply by linking atoms together by their valence on paper. The most dramatic predictions to result from Kekulé's ideas would come from his benzene theory of 1865 that would be celebrated in 1890. Yet despite the power of his theory, Kekulé did not promote his own ideas vigorously, and was relatively unclear and uncertain about how to express graphically the rational formulas of compounds. In a series of papers appearing in 1861, the Russian chemist Aleksandr Butlerov argued that Kekulé's principles would lead to a unique "chemical structure" for each compound that would express all of its chemical properties (Rocke 1981).

Other chemists would quickly develop more graphical methods of notation, the most influential of which was by Alexander Crum Brown, first published in 1864 and still used today. Using straight lines radiating from a central atom to depict the valence, the construction of a formula was simple. Using his novel notation, the two known alcohols with the molecular formula C_3H_8O, for example, could be represented by the formulas in Figure 0.2, in which one alcohol has an oxygen atom at the end of the carbon chain, and the other has an oxygen in the middle.

Crum Brown's formulas have a particularly modern appearance, and when chemists today look at these formulas, without hesitation they would say that they depict a piece of the microworld – a photographic snapshot of a molecule. Yet chemists at the time were extremely reluctant to give this meaning to formulas. Crum Brown, for example, was very clear that his notation meant not "the physical, but merely the chemical position of the atoms" (Brown 1864: 708). This qualification about the meaning of chemical formulas is found throughout the chemical literature of the 1860s, making it clear that chemical structures were not meant originally to be literal pictures of the microworld. This interpretation was closely related to chemists' long-standing commitment to chemical (as opposed to physical) atomism.

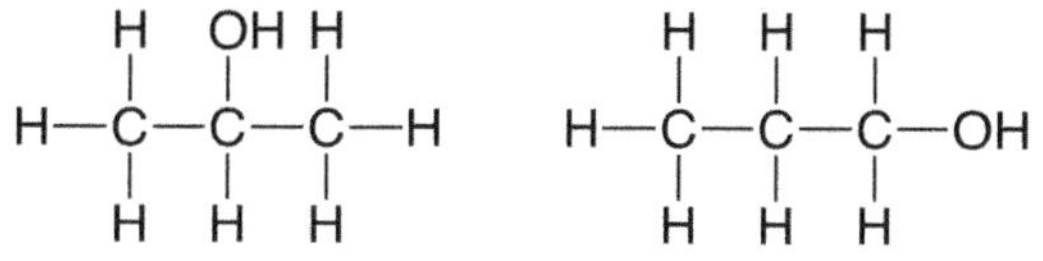

$$\text{H}-\underset{\underset{\text{H}}{|}}{\overset{\overset{\text{H}}{|}}{\text{C}}}-\underset{\underset{\text{H}}{|}}{\overset{\overset{\text{OH}}{|}}{\text{C}}}-\underset{\underset{\text{H}}{|}}{\overset{\overset{\text{H}}{|}}{\text{C}}}-\text{H} \qquad \text{H}-\underset{\underset{\text{H}}{|}}{\overset{\overset{\text{H}}{|}}{\text{C}}}-\underset{\underset{\text{H}}{|}}{\overset{\overset{\text{H}}{|}}{\text{C}}}-\underset{\underset{\text{H}}{|}}{\overset{\overset{\text{H}}{|}}{\text{C}}}-\text{OH}$$

FIGURE 0.2 Crum Brown's notation indicating the difference between two isomers of propyl alcohol. Drawing by Peter J. Ramberg.

Despite this epistemological caution, in 1874, Jacobus Henricus van 't Hoff argued that structural formulas did represent the physical appearance of the molecule in some important ways. In 1848 Louis Pasteur had already noted an association between molecular asymmetry and optical activity (the ability of a certain substances to rotate the plane of polarized light that is passed through them). Van 't Hoff noted that if a carbon atom had four different atoms or molecular groups attached to it in the most symmetrical fashion possible (a tetrahedral array), it would exist in two different asymmetric forms, each form being optically active in opposite senses. Van 't Hoff's claim lay squarely within the tradition of organic chemistry of the nineteenth century: an explicitly *spatial* arrangement would explain specific cases of isomerism that the structural theory alone could not. Although van 't Hoff gave Crum Brown's formulas a meaning chemists had not intended, there was very little serious opposition to his theory, nor did chemists mention that he had crossed an epistemological divide.

The tetrahedral carbon atom proved to be a productive theory, used to explain the instability of the polyacetylenes (Adolf Baeyer) and the isomerism in compounds with double bonds (Johannes Wislicenus) and nitrogen atoms (Alfred Werner and Arthur Hantzsch). Between 1885 and 1892, Emil Fischer provided one of the most dramatic and famous demonstrations of van 't Hoff's theory by differentiating all sixteen possible spatial isomers of glucose and suggesting that the asymmetry of the sugar molecules needed to fit the asymmetry of an enzyme as a key fits into a lock. In 1910, Werner's student Victor King isolated the first asymmetric inorganic compound, and in 1914 Werner had made an optically active compound that contained no carbon atoms, vindicating the idea that asymmetry of the molecule alone, and not its chemical composition, was responsible for optical activity (Ramberg 2003).

A crucial component of in the emergence and success of organic chemistry was the development of a unique visual culture that required the ability to switch seamlessly between paper, hand, and imagination. The first of these visual languages was the Berzelian notation that greatly facilitated writing and printing formulas and offered simple, algebraic-like representation of organic compounds that allowed chemists to "balance" reactions according to the principle of the conservation of weight. Berzelian formulas allowed chemists to "discover," by manipulation of formulas on paper, the constitution of the products of a reaction, and the radicals within those products – Liebig and Wöhler had "discovered" the benzoyl radical as regularities within the Berzelian formulas on paper (Klein 2003).

Throughout the nineteenth century, these "paper tools" were extended to both graphical representations of bonding in the molecules (Crum Brown's stick formulas) and hand-held three-dimensional models such as balls and wires or van 't Hoff's cardboard tetrahedra. Chemists developed a sophisticated array of techniques involving the visualization of chemical structures on paper, in hand-held models, and in their own minds, switching effortlessly between all three (Klein 2003; Ramberg 2003; Meinel 2004; Rocke 2010; Ramberg 2014). All of these methods allowed chemists to predict the existence of possible new compounds and to suggest ways of making them by analogy to known structurally similar compounds. This plurality of methods and the ability to visualize with pencil, hands, and mind is today (with the addition of computer models) still an essential component of chemical culture and practice.

Another factor in the developing culture of organic chemistry was the multifaceted role played by the synthesis of new compounds. A complex term in chemistry with multiple meanings, a "synthesis" could be haphazard or accidental, or a deliberately planned "total" synthesis from elementary materials. Synthesis could mean a simple preparation of a substance following a recipe, or it could mean the opposite of analysis, the "building up" of a complex substance from simpler, elementary components. This last type of synthesis emerged in Liebig's laboratory during the 1840s, when August Wilhelm Hofmann developed "synthetical" methods, or the application of particular chemical reagents to elucidate the constitution of unknown compounds. These synthetic methods were developed specifically to unravel the complex constitution of the newly discovered alkaloids that had proven difficult to understand using combustion analysis alone (Jackson 2014b). Such syntheses were never deliberate or planned, but synthesis by analogy would prove extraordinarily productive, and remains so today.

Friedrich Wöhler's 1828 accidental synthesis of urea from ammonium cyanate is perhaps the most famous of all syntheses of the nineteenth century. Because urea was produced by living things, Wöhler's synthesis from inorganic materials has been celebrated as the defeat of vitalism. At the time, Wöhler and Berzelius both noted the implications of the synthesis for vitalism, but they both already had long been convinced that organic and inorganic compounds should follow the same laws of chemical combination, so the artificial production of urea did not come as a surprise to them. Textbooks of organic chemistry were silent on the matter until the 1840s, and Wöhler, Berzelius, and other chemists primarily saw it as yet another example of isomerism, rather than a synthetic disproof of vitalism. Wöhler's synthesis only took on its later significance for vitalism after Wöhler's death in 1882, in part because German chemists wished to place the origins of the powerful German chemical community, in which synthesis played a central role, squarely in their own country (Brooke 1968; Ramberg 2000).

Many other syntheses of naturally occurring organic compounds would follow, most prominently Hermann Kolbe's syntheses of acetic acid (1845), salicylic acid (1860), and formic acid (1861). Yet none of these syntheses were intentional, but rather were done as part of another research program (Jackson 2014b). In France, Marcellin Berthelot began a systematic program of intentional synthesis in 1854, when he produced hydrocarbons by streaming the inorganic gases carbon disulfide and hydrogen sulfide across hot copper. He made numerous other compounds from inorganic materials with intensive heat, electrical sparking, and other methods. His work culminated in the 1,500-page *La chimie organique fondée sur la synthèse* (1860), where he took great pride in the novelty of his work, paying scant attention to earlier syntheses. For the first time, he described his production of organic from inorganic compounds as "total synthesis." Berthelot saw this program as a vindication of the unity and identity of forces in the organic and inorganic worlds, "establishing the identity of the forces which act in mineral chemistry with those which act in organic chemistry, by showing that the former suffice for producing all the effects and all the compounds from which the latter are born." Uniting the organic and inorganic worlds was an essential part of Berthelot's cultural goal of vindicating his opposition to vitalism and support of a materialist program (Berthelot in Russell 1987; Rocke 2000a: 244–50).

Despite the proliferation of artificial compounds, organic chemistry also remained tied to the study of compounds of plant and animal origin, and as such would be closely related to medicine and physiology. Many plants had been known for centuries to have medicinal, narcotic, or toxic properties – opium poppies from the Middle East, cinchona bark from Peru, or the poisonous seeds from the *Strychnos nux-vomica* tree of Southeast Asia. In the early nineteenth century, chemists began to study these materials, looking for "active principles," the specific substances that caused the physiological effects of these plants. In 1817, the pharmacist Friedrich Wilhelm Sertürner identified morphine as the active principle in the opium poppy. To test it, Sertürner and three volunteers ingested morphine and found it caused drowsiness, palpitations, and stomach pain in such force that they almost immediately induced vomiting to expel the compound. Among its chemical properties, Sertürner also noted that it was basic, the first isolated alkaline plant substance. The French chemists Pierre-Joseph Pelletier and Joseph-Bienaimé Caventou subsequently reported many additional alkaline plant compounds, including brucine and strychnine from

Strychnos nux-vomica (1818), and quinine from cinchona bark (1820). The relatively rapid isolation of these new compounds, all of them alkaline and containing nitrogen, resulted in a new class of organic compounds, named "alkaloids" to complement the already known organic acids. Quinine would soon become extracted commercially from cinchona bark as an effective treatment for malaria, and morphine was adopted for the alleviation of pain (Lesch 1981; Sneader 2005).

By the end of the century, the power of structural theory and synthetic chemistry enabled Paul Ehrlich to study the molecules making up cells. Ehrlich routinely stained microorganisms with coal-tar dyes as a tool for discerning the structure of cells. Some dyes would kill bacteria or stain particular parts of the cell, whereas in other places the same dye might have no effect. Ehrlich realized that the structure of the dyes and where they were absorbed yielded clues about the corresponding structure of the molecules making up the cells, and he began modifying compounds for clinical testing. He found that methylene blue would cure low-grade cases of malaria and arsphenamine (later named Salvarsan) provided the first effective treatment for syphilis (Travis 1989; Ball 2003; Sneader 2005).

In 1828, Joseph Buchner isolated salicylic acid as the active ingredient in willow bark, which had long been known to be effective against fever and inflammation. In 1860, Hermann Kolbe developed a synthesis of salicylic acid from the coal-tar derivative phenol. Large quantities could be produced more efficiently and cheaply using this method, and salicylic acid became a standard treatment for pain and fevers, although it had a tendency to irritate the stomach. In 1896, Felix Hoffmann in the Bayer research laboratory used the newly established technique of acylation – modifying a compound by adding a particular group of carbon and oxygen atoms – to create aspirin, a modified version of salicylic acid that did not cause stomach irritation. Hoffmann also modified morphine using a similar technique to produce heroin (named after its "heroic" power), but unfortunately in this case he produced a far more addictive and dangerous drug (Sherman 2017: 32–4).

While pharmaceuticals remained either the natural products themselves or slightly modified versions, artificial compounds would become commonplace as anesthetics. Early in the century, Humphry Davy's nitrous oxide, or "laughing gas," was used as a dental anesthetic. Inhaling ether had been known since the sixteenth century to have soporific effects, but its general anesthetic qualities were first recognized in the nineteenth century, first tentatively by John Snow during the 1840s, followed by the first public demonstration by William Morton in 1846. Chloroform was first made and characterized in the 1830s, and recognized as an anesthetic in the 1840s by James Young Simpson during a "chloroform frolic party" in which the guests fell asleep after inhaling it. Simpson first used chloroform as an anesthetic in 1847 to relieve pain during childbirth (by Queen Victoria, among others), and it became widespread. Both ether and chloroform were also adopted for use in battlefield surgery, most notably in the American Civil War (Sherman 2017: 41–6).

CREATING A GENERAL CHEMISTRY

By the 1850s, organic chemistry had become the dominant subdiscipline of chemistry, and nearly all academic chairs in chemistry were filled by organic chemists. In their quest to answer questions about composition, arrangement, and structure, however, chemists had largely ignored other questions about the nature of the chemical reactions themselves. Nevertheless, a small number of chemists did research outside the mainstream concerns of composition and structure. Their work fell into three main types. The first was a continuation of the development of analytical techniques for the detection and identification of substances. In Heidelberg, Robert Bunsen maintained the tradition of analytical chemistry, developing instruments for measuring physical properties of chemical compounds, the most famous of which was the spectroscope developed in collaboration with the physicist Gustav Kirchhoff, which they used to identify existing elements and discover new ones (Kirchhoff and Bunsen 1860; McGucken 1969; Hentschel 2002).

The second group studied the relationship between chemical composition and physical properties. From 1839 to 1842, Germain Hess published a series of articles that argued that the heat generated by acid-base reactions was independent of the route used for neutralization. Hess's work was closely linked to Daltonian laws of composition and stoichiometry by considering heat as a reagent in chemical equations (Schelar 1966). In the 1840s, Hermann Kopp carefully measured boiling points of organic compounds to find a direct correlation between boiling point and the number of carbon atoms in the molecule. In the 1860s, Hans Landolt carefully measured refractive indices and the optical activity of organic compounds to correlate these properties with their composition. In 1878, François Raoult described a new method for determining the molecular weights of compounds in solution by freezing point depression.

The third group of chemists was interested in understanding the nature of chemical affinity – why chemical reactions occur in the way they do. Since the eighteenth century, chemists largely had thought about affinity as an attractive force between reacting molecules, a variant of universal gravitation or electrostatic attraction. By mid-century, chemists began to realize that affinity could only be measured indirectly. In the 1850s and 1860s, Julius Thomsen and Marcellin Berthelot independently began extensive research programs looking at the production of heat in chemical reactions, and they concluded independently that the heat evolved in a chemical reaction was an indirect measure of chemical affinity. According to Berthelot and Thomsen, chemical reactions produced products that maximized their affinities with a simultaneous release of heat (Holmes 1962; Servos 1990; Kragh 2016).

Until the 1880s, all of these approaches were accomplished in relative isolation, both professionally and geographically, and had not made use of the new science of thermodynamics, which had been developing in physics from the 1840s and 1850s. During the 1880s, interest in affinities and the application of thermodynamics to chemical reactions would come together in the new physical chemistry created and promoted by three young chemists located geographically outside the mainstream of chemical practice – Wilhelm Ostwald (an ethnic German at the polytechnic in Riga, Latvia), van 't Hoff (in the Netherlands), and Svante Arrhenius (in Sweden). All three advocated a new "physical" chemistry based largely on the chemistry of solutions. Ostwald, the ambitious leader of the movement, saw the continual production of new compounds by organic chemists as uninteresting, and wanted to focus on the reactions themselves to create a truly "general" chemistry that would apply to all chemical reactions. Van 't Hoff followed a reductionist program, striving to mathematize chemistry where possible, and integrate chemistry with physics and mathematics. In *Études de dynamique chimique* (1884), van 't Hoff successfully recast the concept of equilibrium and affinity in terms of thermodynamic principles by defining equilibrium in terms of reaction rates instead of attractions between atoms, and by deriving an equation to show the relationship between temperature and equilibrium. In his doctoral dissertation of 1884, Arrhenius drew on established work in electrochemistry and conductivity to create a new theory of solutions, in which salts were dissociated into ions.

For all three chemists, the central issue became understanding the nature of solutions using Arrhenius' ionic theory. One of the more striking successes of van 't Hoff's approach was the recognition that osmotic pressure of solutions followed an equation entirely analogous to the well-known equation governing gases, but dependent, as Arrhenius pointed out to van 't Hoff, on the number of ions in solution formed by a given salt (Brock 1993). Ostwald, van 't Hoff, and Arrhenius would actively promote their "ionist" chemistry, and founded a new journal, the *Zeitschrift für physikalische Chemie* in 1887.

Of all the ionists, however, Ostwald was the most active proselytizer, through the *Zeitschrift*, numerous textbooks, and above all, creating an active, large, and cosmopolitan research laboratory. In 1887, Ostwald moved from Riga to the larger and more prestigious University of Leipzig to occupy a second chair of chemistry dedicated to physical chemistry. Ostwald attracted students from around the world who would take his principles of "general chemistry" back to their own countries, although the rate of adoption varied. In Britain, aside from William Ramsay, there were few adherents to the ionist program. In France, the new physical chemistry found little traction in the face of Berthelot's opposition in Paris and a relative lack of exchange of students with Germany. Chemists in the French provinces were more receptive, however (Nye 1986). German universities were dominated by the organic chemists who maintained the hegemony of organic chemistry. Other than Ostwald's Leipzig institute, only three other universities had chairs dedicated to physical chemistry, but seven of the eleven technical schools in Germany had physical chemists.

At the turn of the twentieth century, physical chemistry would find its greatest success in the growing chemical profession in the United States. At least forty Americans studied with Ostwald, many of whom had become disillusioned with organic chemistry and were excited by the allure of an exotic new field that had promise to unlock the general laws of chemical reactions. During the 1890s, as American chemists returned from Germany, they established flourishing programs in physical chemistry at universities throughout the United States, many of which began to rival programs in Germany. Among the most significant were the laboratories established by Arthur Noyes at MIT, Theodore Richards at Harvard, Wilder Bancroft at Cornell, and Louis Kahlenberg at Wisconsin. The enormous growth of physical chemistry can be attributed in part to the increase in US college enrollments during the 1890s, which put pressure on American universities to educate an increasing number of students in pre-professional programs, particularly in medicine. Courses formerly devoted to analytical chemistry were subsequently converted to courses in general chemistry that emphasized the new physical chemistry. American physical chemists, with their enthusiasm about their ability to teach a truly general chemistry, leapt at these opportunities, at the same time maintaining their research interests in their graduate laboratories. The result was a flourishing physical chemistry community in the United States (Servos 1990).

PEDAGOGY AND PROFESSIONALIZATION

The significantly changing theoretical and practical culture of nineteenth-century chemistry would be accompanied by equally important changes in the training of chemists and the emergence of professionalization. Chemists had long been trained by direct instruction in the laboratory in a master–apprentice relationship, but until the nineteenth century, this transmission of chemical knowledge and practice had remained at a relatively small scale. The enormous increase in the number of chemists during the nineteenth century can be attributed to the success of the chemical teaching–research laboratory at German universities in the decades after the 1830s. This new method of instruction both greatly increased the volume of research results and the number of trained chemists, and shifted the center of major chemical research from France (specifically Paris) to Germany by the end of the century. Chemists in other countries soon followed the lead of German universities, creating their own laboratories for training students in chemistry.

The origins of the teaching–research laboratory lay in the early nineteenth century, when officials in both France and the German states (and later in Britain) showed a new interest in public health, including water purity, food adulteration, and most importantly, the quality of medicines produced in an expanding pharmaceutical marketplace. These reforms led to the training of pharmacists that required specific examinations on the chemical analysis of medicines that would require formal training. In response to these reforms, a number of private chemical–pharmaceutical institutes were founded in the German states for training chemists in analytical techniques. This new demand for state-qualified pharmacists, coupled with the recent introduction of relatively simple analytical tools like the blowpipe, improved balances, and simple chemical tests, resulted in a large increase in chemists trained in analytical techniques (Homburg 1999).

One of these earliest private institutes was founded in Erfurt (Prussia) by Johann Trommsdorff in 1795 and survived until 1828. Trommsdorff trained students closely in drug preparation and analytical techniques using lectures and close laboratory instruction (Holmes 1989b). The number of pharmaceutical institutes increased during the 1820s, after Prussian reforms of the pharmaceutical profession required all pharmacists to demonstrate their competence in chemical analysis. Many of these new institutes formed at German universities and offered lecture courses and practical training in analytical techniques. By the 1830s, the first textbooks devoted to analytical chemistry appeared, and nearly all German universities had chairs in chemistry with associated laboratories.

Among the most influential of these new university laboratories was run by Friedrich Stromeyer at the University of Göttingen, who had already begun teaching optional courses in analytical techniques in 1810 as part of the medical faculty. By the 1820s, Stromeyer was attracting hundreds of students from all over Germany, training them as pharmacists, physicians, agronomists, mineralogists, mining officials, and many of the next generation of university professors (Homburg 1998). Although Stromeyer was certainly among the first chemists to introduce regular laboratory instruction at a university chemistry laboratory, his laboratory would soon be eclipsed by Justus Liebig's laboratory at the provincial Hessian University of Giessen during the 1830s. Liebig had arrived in Giessen in 1824 after two years in Paris, where he had learned sophisticated analytical techniques with Joseph-Louis Guy-Lussac. On his arrival in Giessen, Liebig attempted to create a chemical–pharmaceutical institute similar to Trommsdorff's, and initially ran his institute privately, because chemistry was not generally regarded as sufficiently scholarly for a university and his proposal was rejected by the Faculty Senate. Liebig initially had few students, and his own research lacked focus and originality, in part due to a lack of the specialized equipment that he had used in Paris.

Liebig's fortunes would change by 1830 after he turned more exclusively to organic chemistry and invented a new apparatus, the *Kaliapparat*, for the elemental analysis of organic compounds. This new apparatus produced results equal in accuracy to previous methods, but was so simple that Liebig's students could learn the technique quickly, and enabled the growth of Liebig's laboratory during the 1830s. Giessen would become the destination of choice for many aspiring chemists from around the world who wanted to learn the new techniques. By the 1840s, Liebig had as many as sixty-eight students in the laboratory, with a significant number of advanced students engaged in coordinated research projects (Morrell 1972; Holmes 1989b; Rocke 2003).

By the end of the century, Liebig's *Kaliapparat* and new pedagogical model had become the standard in teaching chemistry, spread around the world. In Liebig's Germany, the teaching–research laboratory spread quickly. Friedrich Wöhler, a close friend and colleague of Liebig, adopted both the *Kaliapparat* and his pedagogical model at the University of Göttingen by 1838. In 1840, Robert Bunsen initiated a laboratory course in Marburg modeled after Liebig, and continued the practice after he transferred to Breslau and then Heidelberg. At roughly the same time (1840), Otto Erdmann founded a similar approach in Leipzig and Carl Löwig created one of the first teaching–research laboratories outside of Germany at the University of Zürich (Rocke 2000b; Ramberg 2015).

All of these early German laboratories were modest, and were often housed in small, repurposed buildings that quickly filled to capacity. The cost of running the laboratories was also initially modest, and although each of the professors was given a small laboratory budget, it was often not sufficient to cover the total cost of running the laboratory. This began to change by the 1860s, when the German states, spurred by crop failures and Liebig's vigorous promotion of applied chemistry, began a wave of building large, expensive chemical institutes to foster chemical research, accommodate the influx of students, and attract prominent faculty (Borscheid 1976; Rocke 2000a). The state took on more costs of running the laboratory, and the number of faculty also increased. As a sign that interest in organic chemistry had surpassed analytical and inorganic chemistry, the directors of most of these institutes were organic chemists. These large, second-generation institutes in turn became inadequate by the 1890s as chemistry became increasingly fragmented into subdisciplines of inorganic, physical, biological, and radiochemistry, with increasingly divergent methods and aims. Until the end of the century, the apparent unity of chemistry as a science had justified the appointment of a single professor, but the emerging subdisciplines of chemistry put a strain on this hierarchical system. Already in Leipzig and Berlin, "second" chemical institutes had been established, which were largely devoted to physical chemistry (Johnson 1985; Johnson 1989).

In France, Jean-Baptiste Dumas and other chemists almost immediately adopted the *Kaliapparat*, but partly for political and sociocultural reasons,

French chemists were unable to replicate Liebig's teaching–research model. Dumas pleaded with authorities for funding to recreate the conditions of Liebig's Giessen laboratory in Paris, arguing that it is not necessary "to wait for a question to be resolved by the individual work of one of the [faculty's] professors extended over several years, when it can do so in a few weeks under his direction by the collective effort of a dozen beginners in science … " (Dumas in Rocke 2003: 112). Later generations of chemists, including Adolphe Wurtz and others who had experienced Liebig's laboratory first-hand, made similar appeals without success, and unlike laboratories at German universities, most French university laboratories remained small and undernourished institutions, some of them privately funded (Rocke 2000a; Rocke 2003).

Chemical laboratories in Great Britain began to appear in the 1840s, when numerous British students returned from extended periods of study in Liebig's Giessen laboratory. Liebig's visits to Britain between 1837 and 1844 contributed to a growing perception that Liebig's laboratory, with its practical applications of chemistry, had made the Germans superior in agriculture and industry. These events laid the groundwork for new chemical teaching laboratories at the Pharmaceutical Society (1844) and the Birkbeck Laboratory at University College, London (1846). The most famous of the new laboratories would be the Royal College of Chemistry (RCC), established in 1845, with the close consultation by Liebig and the support of Prince Albert. On Liebig's recommendation, his student August Wilhelm Hofmann was appointed the first instructor of chemistry at the RCC. Hofmann followed Liebig's model closely by considering intensive instruction in chemistry as a serious intellectual endeavor that could also be applied to practical problems (Roberts 1976b; Rocke 2000b; Jackson 2006). Throughout the nineteenth century, chemical instruction at Oxford and Cambridge would languish relative to these London schools, but the creation of Owens College, Manchester in 1851 would result, under the direction of Henry Roscoe, in a flourishing center of chemical education and research (Russell 1996; Morris 2015).

The first attempt at creating the teaching–research laboratory in the United States came in 1847, when Eben Horsford returned from Liebig's laboratory with a call to Harvard University. Harvard's president had managed to acquire a substantial gift for establishing a chemical laboratory, but high construction costs and the difficulty in acquiring state of the art equipment in America meant that there were insufficient resources to adequately build and maintain the facility. Horsford managed to attract a steady but small number of students, but was unable to provide full training, and students from his laboratory would still need go to Germany to complete their training in chemistry (Rossiter 1975; Rocke 2003). During the 1850s, Charles Joy at Union College in New York also attempted, at great expense, to create a laboratory explicitly modeled on German laboratories, although in the long term it proved unsuccessful. In 1857,

Joy moved to Columbia University, where he and his student Charles Chandler would create a large and more successful teaching laboratory in connection with the newly established Columbia School of Mines (Bogert 1931; Morris 2015: 186–90).

In 1876, Ira Remsen would initiate a more successful attempt at creating a research laboratory at the newly established Johns Hopkins University. Remsen had studied chemistry in Germany and Daniel Coit Gilman picked him to establish a research program at Johns Hopkins, which Gilman had explicitly modeled after the German research universities. Remsen established a modest but successful laboratory, with a research program in the synthesis and characterization of new organic compounds. By the 1890s, however, Remsen's research and the laboratory had become dated, in part because of his romantic vision of the ill-equipped but successful laboratories of Berzelius, Liebig and Wöhler–Remsen saw the leader of the laboratory as more important than its equipment (Hannaway 1976). As Remsen's laboratory declined, other, far more dynamic and productive laboratories would appear at other universities in the United States, dominated by physical chemists.

Before the 1850s, chemistry in Russia was largely unaffected by the developments in Western Europe. Several universities had been established under Alexander I, the two most important at Kazan (1804) and St. Petersburg (1819). Russian universities appointed chemists to the faculty, but these chemists were usually trained in Western Europe, and once they were appointed to a Russian university, there was little emphasis on original research. There was no tradition of training Russian chemists within Russia, and laboratories were intended for lecture demonstrations and not for hands-on teaching. One exception was Nikolai Zinin at Kazan, who did original and influential research in organic chemistry during the 1840s, although he did not form a large research school. Zinin's student Aleksandr Butlerov became enthusiastic about chemical research while in France and returned to Kazan, where he expanded the teaching laboratory, emphasized original research, training students as chemists within Russia. Butlerov and his successors in Kazan would create perhaps the most successful program in organic chemistry in provincial Russia (Lewis 1994a; Lewis 1994b; Lewis 2012).

By the 1860s, St. Petersburg would compete with Kazan as the growing center of Russian chemistry. Already in 1857, N.N. Sokolov and A.N. Engel'gardt had started a Russian language chemical journal and a private laboratory, strongly influenced by Liebig's model, in St. Petersburg, but both of these enterprises had failed by 1860 (Brooks 1995). During the following decade, a contingent of younger Russian chemists returned to St. Petersburg from studying abroad, many of them with Emil Erlenmeyer in Heidelberg. While in Heidelberg, they formed a *kruzhkók* ("circle"), a well-established social unit developed among the Russian intelligentsia for cultivating and spreading new knowledge. This close intellectual group was then transplanted into St. Petersburg, forming the nucleus of the new Russian Chemical Society in 1868. Butlerov's move from Kazan to St. Petersburg in the same year would accelerate the increasing professionalization of Russian chemists (Graham 1993; Brooks 1998a; Brooks 1998b; Gordin 2008).

NATIONAL STYLES AND INTERNATIONALISM

The theoretical and pedagogical culture of chemistry during the nineteenth century was also significantly affected by broader political and cultural movements. In the struggle between European nations for political, economic, and colonial hegemony, chemistry became a crucial source of material wealth and general prosperity. International rivalries between nation-states were of a partly chemical nature, from the control of raw materials, the production of chemical goods, the production of synthetic compounds, and the academic scientific prestige of chemical institutions. The rise of the nation-state and nationalism resulted in the creation of romanticized national heroes, including scientists and inventors, who from humble origins achieved the glory of unprecedented scientific achievements. Many chemists, including Pasteur, Berthelot, Curie, Davy, Dalton, Faraday, Liebig, Kekulé, Mendeleev, and Volta, became national heroes in their respective countries.

However, historians have also moved beyond the narrative of the "heroic" chemist to uncover significant national rivalries and differences in national styles that significantly affected the spread of chemical theories and practice. The different receptions of physical chemistry in the United States and European countries illustrates how broader cultural, political, and institutional factors can affect the practice of chemistry across national boundaries. Another significant difference was between Germany and France. Academic culture in France was centralized in Paris, and French chemists were strongly influenced by Auguste Comte's positivism, leading to a more skeptical view of the atomic theory and the later structural theory of organic chemistry.

The university structure in the German states, on the other hand, was less centralized as a result of the proliferation of teaching–research laboratories based on the Liebig model at the many German universities. German chemists, less constrained by positivism, and more inclined to exploit the new structural chemistry, had the laboratories necessary for the preparation of the new compounds predicted by the theory. At mid-century, French chemists, still dominated by the centralized Parisian model with fewer laboratories, were unable to exploit the new chemistry as well as the Germans. By 1870, chemistry in France had notably declined relative to Germany, both in the number of published articles and in the number of teaching positions (Rocke 2000a: 400–1).

The irony of this story is that the success of the structure theory relied on the reforms advocated by two French chemists, Charles Gerhardt and Auguste

Laurent, during the 1850s. The call for these reforms had fallen on deaf ears in France, while German chemists and Alexander Williamson in Britain had adopted them enthusiastically. Adolphe Wurtz, as a lonely advocate for the reforms in France, knew this all too well, as he attempted to appeal to his fellow countrymen by opening his *Histoire des doctrines chimique* (1869) with the words "chemistry is a French science, founded by Lavoisier in immortal memory" (Rocke 1994).

Despite the rivalries between countries, there were also attempts to improve internationalism and create an agreement on fundamental issues. The Karlsruhe conference of 1860 was organized to reach an agreement on atomic weights (Nye 1984). Another major international convention, the 1892 Geneva Congress on Organic Nomenclature illustrates the limitations of internationalism. Under a supposedly open-minded cosmopolitan atmosphere, the French chemist Charles Friedel campaigned against the supposedly "disordered" German nomenclature, opposing Adolf von Baeyer's proposal for a new systematic nomenclature based on structural formulas (Figure 0.3; Hepler-Smith 2015). Ostwald also favored the internationalization of chemical knowledge, serving as the first president of the short-lived International Association of Chemical Societies in 1911, and

FIGURE 0.3 Charles Friedel (seated, sixth from the left), Cannizzaro (seated, fifth from left), Baeyer (seated, fourth from left) at the Geneva Congress for Reform of Chemical Nomenclature. Photograph, 1892. Courtesy of the Wellcome Collection.

endorsing a constructed language called Ido – similar to Esperanto, but better adapted to scientific terms (Gordin 2015; Fox 2016a: 32–5).

Nevertheless, these attempts at internationalism did not eliminate widespread nationalism, epitomized by the bitter Franco-German rivalry. French defeat in the Franco-Prussian War had already incited French hostility against German chemists to the point that their papers were not accepted in French journals for many years. The hegemony of the German chemical industry made things difficult late in the century, when Britain and France had to admit defeat to the German chemical industry's mass production of artificial dyes and drugs. The struggle between Britain, Germany, and France for the control of raw materials in their respective colonies contributed to the outbreak of World War I, in which chemistry also played a crucial role in terms of national rivalries.

In 1914, many German chemists – Baeyer, Fischer, Ostwald, Paul Ehrlich, Walter Nernst, Richard Willstätter, and Fritz Haber, among others – signed the "Call to the Civilized World," a defense of the German Kaiser's decision to enter the war as a response to a long list of the Allies' lies and calumnies against Germany (Fox 2016a: 47–38). The "Call" ended by declaring German *Kultur* as superior to French *civilisation* and the British Empire: "You who know us, who hitherto together with us have cherished the supreme possession of mankind, to you we appeal: believe us when we say that we shall fight this fight to the finish as a civilized nation, for which the legacy of a Goethe, a Beethoven, a Kant is as sacred as its own hearth and acre." In return, scientists in the Allied countries ostracized German scientists, and the rift between Allied and German scientists would grow deeper with the Allies resentment of Germany's use of chemical weapons during the war. It would only be fully healed well after the war's conclusion (Agar 2012: 114–15; Fox 2016a: 47–8).

Like many other sciences of the time, chemistry could not escape the tensions and paradoxes of nationalistic fervor and the desire for internationalism. Wurtz claimed chemistry as a "French" science, but earlier organized the international Karlsruhe Conference. There were international chemical displays at world fairs, but also fierce versus national rivalries in creating strong chemical industries. Fritz Haber created the process for production of artificial nitrogen fertilizers, but also enthusiastically participated in developing chemical warfare agents for the German army. Chemistry reflected the complex and often contradictory national–international character of nineteenth-century Europe.

GROWTH AND DIVERSIFICATION IN CHEMICAL INDUSTRY

During the first half of the nineteenth century, the chemical industry was dominated by the production of soda (i.e. sodium carbonate), driven by the increased demand for bleached textiles, glass, and paper, using the process developed by Nicholas Leblanc in 1791. The Leblanc process relied on sulfuric acid, which was subsequently produced in large quantities, usually at the same site as soda production to avoid the danger and cost of transporting it in large quantities. Driven by its large textile industry, Britain dominated the alkali industry in Europe, where whole towns would grow under the shadow of the chimneys of large alkali factories (Warren 1988). Charles Tennant in Glasgow and James Muspratt in Liverpool were among the most prosperous manufacturers. Tennant was already a well-established producer of bleaching powder and sulfuric acid when he began the production of soda in 1818, and by the 1840s, Tennant's firm (he died in 1838) was the largest chemical works in Europe. Tennant's main rival would be Muspratt, who began his career by manufacturing prusside of potassium (potassium ferrocyanide, necessary for making the pigment Prussian blue), alum (as a mordant for dyes), and bleaching powder. In 1828, Muspratt began the production of soda using the Leblanc process, and his alkali works soon became nearly as large as Tennant's. Despite being rivals commercially, Tennant and Muspratt were also friends, and concern over the stability of the supplies of sulfur led them to collaborate in the 1830s to acquire leases to sulfur mines in Sicily. These mines were prone to flooding and proved to be an unviable source of sulfur, and Muspratt turned instead to iron pyrites, exploiting the large quantities of this mineral found in Ireland and Wales (Reed 2015: 43–5).

Although the Leblanc process was successful, it was also very inefficient by later standards, as the final product was only 15 percent of the weight of starting materials, and production had an appalling effect on workers and the environment. It released large quantities of gaseous hydrochloric acid into the atmosphere, contaminated productive farmland with solid waste, and the harsh conditions of working with sulfuric and hydrochloric acid created extremely hazardous conditions for workers in the acid and alkali factories. Some of these problems were reduced by the 1860s, after the adoption of William Gossage's technique for capturing the hydrochloric acid before it was vented from the chimneys. The captured acid could then be sold to meet the demand for creating chlorine to bleach textiles (Russell 2000b). The production of soda by the Leblanc process reached its peak by the 1880s; despite its inefficient batch process, and unhealthy working conditions, it had no rivals for large-scale production of soda.

The Leblanc process would soon be challenged, however, by an alternative method developed by the Belgian Ernest Solvay in the 1860s. The Solvay process involved passing carbon dioxide gas through a brine solution saturated with ammonia gas. It was initially difficult to figure out how to manipulate these gases on a such large scale, but once the production problems were solved the process proved far more efficient than Leblanc soda. Solvay soda resulted in far less waste, and far less environmentally damaging waste, than the Leblanc process, and produced alkali continuously rather than inefficiently in batches.

Among the most successful industrialists to adopt the Solvay process was the German–British chemist Ludwig Mond, who in 1871 negotiated terms with Solvay to use his process in Britain. Despite initial difficulties in raising sufficient funds, Mond and his partner John Brunner began building their factory in 1873. They encountered additional difficulties in getting production started, but by 1876 they had begun to break even, and the business grew quickly based on the high quality of the soda produced. By the 1880s, production of soda by Solvay's method had matched that produced by the Leblanc method, and by the 1890s, the Leblanc manufacturers began switching to the more profitable Solvay method. The Solvay process was the first highly technical chemical engineering process; by the turn of the century it had driven Leblanc soda out of business (Haber 1969; Warren 1988; Russell 2000b).

During the last half of the nineteenth century, the large-scale production of inorganic chemicals would be joined by a rapidly growing new sector of the chemical industry that produced dyes and pharmaceuticals derived from organic compounds. The extraction of dyes from natural materials – madder root and indigo, for example – had long been large and extremely lucrative industries, but from the mid-nineteenth century chemists began to develop an increasing number of artificial dyes made of carbon-based compounds. The key to these new dyes would be coal tar, a liquid by-product of heating coal to produce coke for steel mills and coal gas for lighting. Coal tar was largely a noxious waste product with few uses until the 1850s, when it became a starting material for producing many new organic compounds (Russell 2000b: 219–25). In particular, August Wilhelm Hofmann and his students isolated aniline and other nitrogenous compounds, and Hofmann's student Charles Mansfield developed fractional distillation methods for isolating and identifying compounds such as phenol, benzene, xylene, and toluene, which in turn found their own uses as solvents, varnishes, and dry cleaning agents.

One of the first dyes derived from coal tar was the bright yellow compound picric acid made by nitrating phenol, but the real breakthrough for the coal tar dye industry was a much more complex organic compound, a rich purple dye made by Hofmann's young student William Henry Perkin, who produced it accidentally in a failed attempt to make the antimalarial drug quinine from aniline. Perkin noted that cloth took this dye well, and against Hofmann's advice, established his own dye factory financed by his father. After modifying his initial synthesis to produce mauve on a large scale, he brought the dye to market by 1859. Perkin's "Tyrian purple" (which he later renamed mauve) was a sensation, worn in dresses by both Eugenie Bonaparte and Queen Victoria, making mauve a highly desirable color and a fortune for Perkin and his family (Travis 1993).

Perkin's mauve was one of many new artificial colors that appeared during the 1860s, known as the aniline dyes. Among the most popular was magenta (also known as fuchsine or rosaniline), a red color created by François Verguin in Lyon, and aniline blue. It was magenta in particular that sparked the initial explosion in the number of coal-tar dye company startups. The lucrative nature of aniline dyes resulted in the formation of new chemical firms all over Europe, with the earliest successful firms founded in Britain and France. A number of firms were also established early on in Germany: Bayer (1863), Badische Anilin- & Sodafabrik (BASF, 1865), Hoechst (1863), and Aktiengesellschaft für Anilinfabrikation (AGFA, 1873) (Kay-Williams 2013: 140–1).

Although all of these aniline dyes could be made on an industrial scale, they were discovered in a haphazard, accidental way, often by trial and error. This began to change in the late 1860s, when the new structural theory of organic chemistry guided chemists in the replication of natural dyes and in creating new ones (Rocke 2018a). In 1868, Carl Graebe and Carl Lieberman succeeded in making alizarin, the red compound found in madder root, from anthracene, another coal-tar derivative. A modified synthesis of alizarin by Heinrich Caro at BASF would prove to be extremely successful, quickly replacing madder cultivation as the principal source of alizarin. In 1880, Baeyer succeeded in making synthetic indigo in the laboratory. His process was eventually modified and improved for large-scale production of indigo. By 1897, the cost of synthetic indigo dropped to less than natural indigo, leading to the eventual collapse of the large British-controlled indigo plantations in India (Kumar 2012). Although the transition was slow, by the early twentieth century synthetic alizarin and indigo could be made far more inexpensively, in greater amounts, in more consistent and stronger colors, free of the seasonal dependence of plant-based dye extraction.

The growing power of chemists' control over nature would also become evident in the azo dyes, an artificial class of compounds derived from coal tar. First discovered in 1858 by Peter Griess, the first two azo dyes appeared on the market in 1863, but in 1876, Griess was able to make a wide variety of colors by subtly varying the structure of the starting compound, and the German chemist Otto Witt established a specific relationship between the structure of the azo dye molecule and its color. Chemists could now go beyond nature and produce almost any color they desired, and by 1912, azo dyes would make up nearly half the synthetic dyes on the market (Travis 1989; Travis 1993).

German chemical firms were best able to exploit this new chemistry and by 1878 produced 60 percent of all global dyestuffs, a percentage that would increase further over the next twenty years, eclipsing dye production elsewhere in Europe and North America. The reasons for this are complex, but in part, German firms had access to academic professors and their supply of new chemists to staff their own research laboratories dedicated to the development of new dyes. Industrial firms early in the century had hired trained works chemists tasked with analysis and quality control, and the earliest British and

French dye factories of the 1860s had loosely affiliated private laboratories, but they were not fully integrated with the factories, or a part of corporate planning (Homburg 1992).

By contrast, German firms incorporated research and development into their corporate plans, although the path by which this took place was somewhat unplanned and haphazard. Heinrich Caro was hired by BASF in 1868 to oversee production, not create novel compounds. Hoechst created an analytical laboratory in 1870 for works chemists. A decisive turning point for the development of azo dyes appears to be the German patent law of 1877 that provided an incentive to create new dyes, rather than simply making the production costs of existing dyes more efficient, and corporate management began taking active steps towards creating research laboratories. Caro's research laboratory was fully integrated into BASF's corporate structure in 1877. Hoechst began research on azo dyes in 1878, opened a pharmaceutical laboratory in 1883, and a central research laboratory in 1893 as the deliberate result of plans by management for developing new production. AGFA created a research laboratory in 1882, and all three companies actively recruited academic chemists to direct the laboratories (Homburg 1992; Reinhardt and Travis 2000). Bayer was slower at establishing a central research facility. Carl Duisburg had been hired in 1884 to oversee azo dye production, and had accumulated a growing number of assistants developing new dyes and insuring their patents, but their research was scattered over many locations overseeing various production processes with no authoritative coordinating oversight (Meyer-Thurow 1982). Finally, in 1889, Duisburg convinced Bayer to build a dedicated central research facility that was completed in 1891. All of these industrial laboratories were modeled after existing academic research laboratories, although they were somewhat more crowded and less elaborate (Morris 2015).

The success of all dye firms relied on actively developing markets by creating a demand for new colors, and the new inexpensive dyes played a role in the democratization of fashion of the late nineteenth century. Until this time, natural dyes were expensive, making colorful clothes affordable only to the prosperous. Lower-class women were able to afford only plain clothing with dull colors, or second-hand out-of-fashion recut clothing. During the last half of the nineteenth century, a number of innovations made fashionable clothing more broadly available, including the mass production of textile materials, treadle sewing machines, printed paper patterns for sewing clothing, fashion magazines (that often provided paper patterns), and department stores that were licensed to make copies of the most popular designs using fabrics they also sold in-house. While all of these factors were significant, the new synthetic dyes also played an important role. The difficult patterns for expensive clothing might be hard to replicate, but colors provided an easy way to make articles of clothing stand out. Colors could also be changed quickly to create a cyclical demand.

The new dyes were also inexpensive, making dyed fabrics more available and affordable, increasing their popularity among all classes (Figure 0.4; Forster and Christie 2013).

At the same time the dye industry was emerging, metallurgy would be transformed by new methods for producing large quantities of inexpensive high-grade steel and aluminum. Steel was initially difficult to produce inexpensively in large quantities. The first efficient production of steel was developed by Henry Bessemer, who invented a process for removing carbon and other impurities from molten iron by passing heated air through the molten iron. Bessemer's process greatly reduced the cost of steel, enabling construction of steel bridges, ships, railroads, buildings, and munitions. Sydney Gilchrist Thomas improved upon Bessemer's process by inventing a method for removing phosphorus from low-grade iron ore. The Gilchrist-Thomas process would prove invaluable to the German steel industry, allowing the full exploitation of high-phosphorus iron ore in the Lorraine region, annexed by Germany after the Franco-Prussian war, and rapidly increasing Germany's output of steel (Hendersen 2006). The Bessemer process would eventually be overtaken in the early twentieth century by the Siemens–Martin open-hearth process that created the high temperatures

SCENE—COMMERCIAL ROOM.

INCIPIENT COMMERCIAL TO CRUSTY OLD TRAVELLER. "*You're always in the Fashion, I see. Last time I had the pleasure of seeing you, Mauve was the prevailing Colour, and your Nose was Mauve. Now Magenta is all the go, and it's changed to Magenta.*"

FIGURE 0.4 An 1861 cartoon that comments on the spectacular success of the new synthetic dyes (*Punch*, November 23, 1861: 212). Courtesy of the Sidney M. Edelstein Center, The Hebrew University of Jerusalem, Israel.

for smelting. The process was slower, but produced larger quantities of steel more efficiently, created more control over the specific content of the steel, and allowed recycling of scrap iron for the first time.

Late in the century, the production of inexpensive metallic aluminum introduced large quantities of an entirely new metal for industrial and commercial use. Aluminum is among the most plentiful of elements in the Earth's crust and is commonplace today, but for many years it was exceedingly rare and expensive to produce in its metallic form. As a young student at Oberlin College in Ohio, Charles Martin Hall learned about the properties and rarity of aluminum from the professor of chemistry, Frank Jewett. Hall and Jewett collaborated on finding more practical methods for isolating aluminum metal, eventually creating batteries that produced electrical current strong enough to isolate small quantities of aluminum metal electrochemically. After graduation, Hall created a workshop behind his family's house to improve this process, and in collaboration with his sister Julia found that passing a strong electrical current through aluminum oxide dissolved in molten cryolite (sodium aluminum fluoride) at 1000 °C produced large quantities of metallic aluminum. Although Hall's application for a patent was challenged by Paul L.T. Héroult in France, who had developed a similar process nearly simultaneously, Hall managed to demonstrate that he had developed his process before Héroult, and retained the US Patent. As a lightweight but strong metal, aluminum would eventually find innumerable industrial and household applications, and Hall's commercialization of the process would lead to the formation of the Aluminum Company of America (Alcoa; Craig 1986; American Chemical Society 1997).

ATOMISM REDEFINED

In late 1895, Wilhelm Röntgen published his experiments on mysterious "x-rays," generated from an evacuated cathode ray tube, that could penetrate matter and take pictures of bones in living tissue. Soon afterwards, Henri Becquerel noticed that uranium salts wrapped in thick paper would create an image of the mineral on photographic plates. These "Becquerel Rays" generated by uranium were soon found to have an electrical charge and that they could create charged ions in gases. In 1897, Marie Skłodowska Curie, an immigrant from Poland trained in Paris, began her famous study of uranium and other minerals that exhibited this new "radioactivity," a term that she coined in 1898. In the same year, she and her husband Pierre announced, after months of arduous labor, that uranium ore contained two new elements whose radioactive power was far greater than uranium itself: polonium and radium. The nature of x-rays and radioactivity would prove puzzling to both physicists and chemists, but would also become a central component of the new physics and chemistry of the early twentieth century, and an essential tool for understanding subatomic structure (Nye 1996: 147–53).

A bewildering variety of radioactive elements would follow, and untangling the relationships between these new elements would require a combination of physical methods to quantify the different types of emanations and chemical methods to separate and identify each species. By 1900, the emanations had been separated into three different types – simply named α-, β-, and γ-radiation, with relative penetrating power in that order – and numerous experiments also suggested that radioactivity arose from within the atom itself (Nye 1996: 154).

In 1903, after a careful, painstaking examination of a long chain of emanation products beginning with radium and thorium, Ernest Rutherford and his collaborator Frederick Soddy at McGill University in Montreal concluded that radioactive atoms were inherently unstable. These atoms emitted particles that both changed its atomic weight and changed it into a new element. Radium, for example, lost an α-particle to become radon, a new gas identified by Rutherford and Soddy; radon in turn lost an α-particle to become polonium, and polonium turned into non-radioactive lead. In general, the radioactive elements transmuted into atoms of different elements (Figure 0.5; Nye 1996: 156).

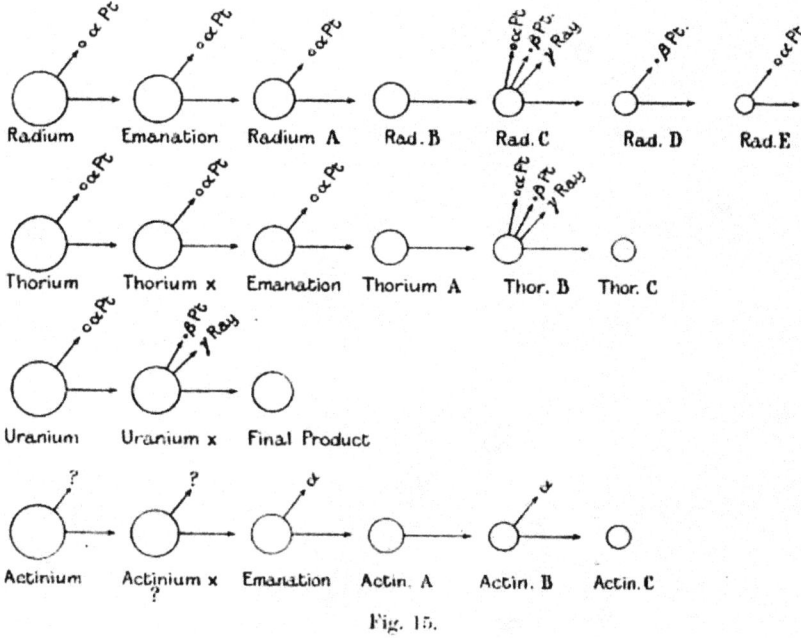

FIGURE 0.5 Rutherford's illustration of transmution of elements by α-, β-, and γ- radiation (Rutherford 1906). Courtesy of the Oesper Collection, University of Cincinatti.

Rutherford and Soddy suggested that radium itself must be produced from uranium or thorium, because "the radium present in a mineral has not been in existence as long as the mineral itself, but is being continually produced by radioactive change" (Rutherford and Soddy, 1903 in Romer 1964). The Curies and other French physicists were initially skeptical of the transmutation theory, as it became known, until Soddy and Ramsay showed conclusively by spectroscopy that the α-radiation given off by radium bromide consisted of helium (Malley 1979; Malley 2011).

The study of the causes of radioactivity coincided with numerous experiments in physics on the structure of the atom. Experiments with cathode ray tubes, most notably J.J. Thomson's discovery of the electron in 1897, led to speculation about subatomic structure. In Rutherford's hands, alpha particles became a useful tool for probing atomic structure, and in a now famous experiment, Rutherford directed alpha particles towards a thin sheet of gold foil, finding that only a very small number (about 1 in 20,000) were deflected from their original path. The result led Rutherford to conclude that gold atoms were nearly all empty space, and he suggested that atoms contained a positively charged nucleus that made up only a small fraction of its volume (Nye 1996: 164–6).

Rutherford's nuclear model of the atom raised serious questions about how the negatively charged electrons could orbit the very small, positively charged nucleus without falling into it. This problem would be solved by Niels Bohr in 1913, who visited Rutherford's laboratory in 1911, shortly after the gold foil experiment. Bohr made use of the quantum theory, introduced by the physicist Max Planck in 1901, in which energy exists only in discrete portions, and combined it with Rutherford's nuclear model. With a detailed mathematical analysis that assumed electrons were point masses that orbited the nucleus, Bohr argued that electrons could occupy only the discrete energy levels suggested by spectroscopic measurements. Although physicists would revise the model substantially in the 1920s, Bohr's fundamental idea of electrons occupying discrete energy levels would be sufficient for most chemists in the twentieth century (Nye 1996: 167–70).

Two other developments would further refine the definition of elements. First, Henry Moseley, working in Rutherford's laboratory, found in 1913 that each element showed a specific correlation between its x-ray emission and a new parameter that he called the atomic number. Moseley's number would come to replace the atomic weight as the defining characteristic of each element, and clarified the exact number of possible elements in the periodic table (Nye 1996: 149). Moseley volunteered for service in the British Army at the outbreak of war in 1914, and was tragically killed in the Gallipoli campaign a year later at the age of only twenty-seven, a loss that deeply affected Rutherford and other scientists (MacLeod et al. 2018). Also in 1913, Soddy would coin a new term that would redefine the nature of an element. Because several radioactive elements had different atomic weights but identical chemical properties, Soddy suggested that these compounds be called isotopes, and the fractional atomic weights that had puzzled chemists for so long became clear, as they were now seen to be statistically average weights of atoms with the same atomic number, but different integral atomic weights.

In short, between 1895 and 1913, chemists and physicists became convinced that Dalton's view of atoms as solid, homogeneous spheres had become untenable. Rutherford and Soddy had shown that atoms of some elements were not stable, and decomposed into different elements. Atoms were, furthermore, found to consist of particles that were still more fundamental – electrons and the positively charged nucleus (later shown to be made of protons and neutrons). Finally, Moseley and Soddy's research suggested that the atomic weight was not the property that defined an element or its position in the periodic table.

Solving the problem of radioactivity and atomic structure would significantly blur the line between chemistry and physics because the many new radioactive elements could be identified by the physical characteristics of their emanations or their similarity in chemical properties to existing elements. The Curies and Henri Becquerel were jointly awarded the 1903 Nobel Prize in physics, but Rutherford, much to his surprise, was awarded the 1908 Nobel Prize in *chemistry* for the discovery of transmutation. Marie Curie received the 1911 prize in chemistry for the isolation of radium and polonium, and Soddy was awarded the 1921 prize in chemistry for his work on isotopes. The Nobel prizes awarded for chemistry reflected the strong chemical character of work in radiochemistry, entailing the chemical identification of new species and the realization that several varieties of the large number of new radioactive elements had similar chemical properties, and that they should therefore be grouped together as varieties of the same element (isotopes). Yet the connections to physics in the study of radioactivity remained strong, and by the 1920s, subatomic structure would increasingly become the domain of physics.

The discovery of the electron and subatomic structure would also change how chemists thought about the concept of bonding. Although chemists had assumed that atoms were held together in the structure of the molecules, they had not attempted seriously to explain precisely what held them together, although they had long been convinced that it was some sort of electrical force. This changed quickly with the discovery of the electron in 1897 by J.J. Thomson. In the decade after Thomson's discovery, physicists concentrated on incorporating the electron into models of the atom that would explain the spectral properties of elements, but Thomson himself suggested in 1904 that the chemical bond might involve the exchange of a single electron between atoms (Stranges 1982).

With their descriptive and qualitative understanding of molecules, most chemists were at first hesitant to follow Thomson's lead and create models of bonding involving electrons, although tentative theories began to appear in chemistry textbooks as early as 1904. Of the several models developed between

1900 and 1910, the most famous and influential was Richard Abegg's concept of "electroaffinity" or the tendency of atoms to gain or lose electrons based on their position in the periodic table. All early models of bonding assumed a complete transfer of a single electron from one atom to another, resulting in a simple electrostatic attraction between atoms in all molecules. The various electrostatic models would dominate chemists' discussion of bonding until 1916, when Gilbert N. Lewis published his first paper on the electron-pair bond. Although Lewis' ideas resembled Abegg's model, Lewis likely conceived the idea independently, as his initial ideas were contained in a 1904 notebook that predates Abegg's publications. As Lewis further developed his model in the years after 1916, it would come to dominate chemists' concept of bonding for the remainder of the twentieth century (Stranges 1982).

THE BROADER IMPACT OF RADIOACTIVITY

The discovery of the mysterious new phenomena of x-rays and radioactivity had a great impact on the public imagination. The initial announcement of x-rays immediately resulted in a wave of popular and often inconsistent accounts. The rays were thought to be either electrical or they were light waves, or they were a new version of photography. By some accounts they were deadly to living things, by others they possessed vital powers (Lavine 2013: 31–2). The announcement of radium's mysterious powers had an even greater impact on the public, and during the first decade of the twentieth century, the public was fascinated by radium and its mysterious rays, its rarity and extravagant cost, its ability to spontaneously generate heat and glow in the dark, and by the romanticized depictions of Marie Curie's search for polonium and radium as a woman in a field dominated by men (Malley 2011: 209). The use of biological metaphors – half-lives, decay, transmutation – to describe radioactive phenomena encouraged and reinforced the notion that radioactivity, and radium especially, was a source of vitality, perhaps even alive itself (Campos 2007; Campos 2015). Sensational news reports discussed radium's curative powers or its ability to restore sight, and radium became an additive in ointments, medicaments, and dietary supplements (Figure 0.6). In some circles, its mysterious rays were somehow connected to the spiritual world or similar to the effect of the Holy Spirit on Christians (Lavine 2013: 38–9).

The notion that radium had a vitalizing power with the ability to transmute itself also took hold in biology. In 1905, John Butler Burke interpreted Soddy's use of biological metaphors to mean that radium had a vitalizing power, and he attempted to create new life by immersing radium salts in sterilized beef broth. Rutherford's idea that the early Earth would have been highly radioactive suggested that radioactivity might be the original cause of life itself. George Darwin, son of Charles, made an explicit analogy between transmutation of

FIGURE 0.6 Advertisement for radium-containing medicinal products, 1916. Courtesy of the Wellcome Collection.

the elements and the origin of biological organisms by transmutation of species. Just as there were stable and unstable elements, there were also stable life forms (species) and unstable life forms (mutants). Perhaps there could be more than a metaphorical relationship – could a transmuting element be used to cause a transmutation in species? In 1908, C. Stuart Gager attempted to treat plants with radium and concluded that its rays "act as a stimulus to living protoplasm." The famous geneticist Thomas H. Morgan successfully created mutant fruit flies by treating them with radium, and Hermann J. Muller initially used radium to attempt the artificial "transmutation" of genes before he turned to x-rays and the work that made him famous (Campos 2015).

The discovery of radioactivity would also have a significant impact on geology. It became known early on that radium emitted a constant amount of heat, and almost immediately Rutherford and others suggested that this unexpected heat source could extend the estimated age of the Earth calculated from cooling data, which, according to Lord Kelvin's best mid-nineteenth-century estimates, could be no older than a few hundred million years. Rutherford realized that decay rates of radioactive isotopes would allow a calculation of the age of the Earth, and after initially promising but inconclusive work by Bertram Boltwood and Robert Strutt (later 4th Baron Rayleigh), Strutt's student Arthur Holmes would successfully use radioactive dating to establish a much older age for the Earth that would eventually expand the age to several billion years (Burchfield 1975).

Another effect of radioactivity and transmutation on the public was the fear of economic collapse, either by the production of large quantities of gold, or by making gold worthless by destroying it with radiation. These fears emerged as tropes in early science fiction that dealt with the possible consequences of removing the gold standard. In Garrett P. Serviss' *The Moon Metal* (1900), for example, the scientist Max Syx introduces a new rare metal, artemesium, mined by "cathode ray transfer" from the moon, to control monetary stability after an abundance of gold was found in Antarctica. While transmutation did not play a direct role in Serviss' novel, H.G. Wells was directly inspired by Soddy's thinking about the nature of wealth in the light of transmutation to write the prescient novel *The World Set Free* (1914). In Wells' vision of the future, gold becomes worthless as a plentiful by-product of new forms of atomic energy, and Wells imagines the consequences of plentiful gold, personal flying machines, machine labor, and bombs with unimaginable atomic power that required state control of atomic weapons (Morrisson 2007).

Finally, the discovery of transmutation coincided and resonated strongly with the peak of the "alchemical revival" of the late nineteenth century. This revival was initiated by Mary Anne Atwood's *A Suggestive Inquiry into the Hermetic Mystery* (1850), in which she argued that the alchemist must reach a higher mental state to produce the philosophers' stone from the "ether" or the electrical fluid. In *Remarks upon Alchymists* (1855), Ethan Allen Hitchcock offered a similar interpretation of alchemy in strictly religious terms, and Atwood's and Hitchcock's books are the origin of the idea, still popular today but rejected by historians, that alchemy had nothing to do with chemistry and the laboratory, but was about the spiritual perfection of the alchemist. Other movements, such as the formation of the Theosophical Society (1875) and the Hermetic Order of the Golden Dawn (1888), incorporated alchemical ideas, in particular transmutation. In Britain, the occult revival spurred translation of alchemical sources, attempts at transmutation, and created an overall sympathy towards the aims of alchemy, both scientific and religious (Principe 2012: 94–102).

Chemists, too, had become more aware of alchemy and transmutation as part of their distant past. Hermann Kopp, Ernst von Meyer, and Marcellin Berthelot had written extensive histories of chemistry that were read widely by chemists. Some of the most widely read histories of chemistry in Britain were by M.M. Patterson Muir, who published at the height of the alchemical revival of the 1890s. By the 1890s, chemists were giving regular lectures in the classroom and public settings on alchemical topics. Frederick Soddy himself was steeped in alchemy in the classroom, having read the histories by Kopp and Meyer at Oxford, and then later reading Berthelot's sympathetic *L'origine d'alchemie* (1885) that he used to craft his own lectures. William Ramsay was not a practicing occultist, but he was a member of the Society for Psychical Research, and included extensive alchemical topics and documents in his own lectures (Morrisson 2007: 101).

Despite this strong interest in alchemy, nineteenth-century chemists had nevertheless nearly unanimously regarded transmutation as impossible. Muir regarded the "modern alchemist" as simply a "theosophist or theologian," explaining phenomena by an appeal to a "higher spiritual plane" (Muir in Morrisson 2007: 101). However, chemists were also aware, even if they rejected it, of the concept of the underlying unity of matter, the *prima materia*, from the premises of transmutational alchemy, and the legacy of Prout's hypothesis. Soddy and Rutherford's announcement of transmutation in 1903 therefore fell on fertile ground and appeared to vindicate the alchemical dream of transmutation – perhaps, some thought, the alchemists had possessed long-forgotten knowledge. The nature of atoms was now open to dispute, and perhaps the elements themselves could be created in the same way as compounds. In 1911, William Ramsay, quoting Michael Faraday, wrote: "to decompose the metals, to reform them, and to realize the once absurd notion of transformation – these are the problems now given to the chemist for solution" (Morrisson 2007: 110).

CONCLUSION

During the nineteenth century, chemists established sophisticated cultures of intellectual inquiry and professional discipline formation. They established the fundamental ideas of modern chemistry that would endure throughout the twentieth century: atomism, periodicity, composition, structure, bonding, and equilibrium. Like their counterparts in physics and biology, nineteenth-century chemists were preoccupied with the unification of their discipline. Chemists were convinced that organic and inorganic chemistry should follow the same natural laws, and that using analogical reasoning from the organic to the inorganic realms (or vice versa) could help to determine rational formulas. The extraordinarily diverse number of organic compounds could be brought under the general umbrella of "chemical structure." The pervading interest in organic synthesis, including Wöhler's synthesis of urea (and the later mythical significance of that synthesis), Kolbe's synthesis of acetic acid, and Berthelot's synthesis of organic compounds reflected the desire to show that inorganic and organic compounds followed the same chemical laws. All the known elements, and those discovered after 1869, could be placed in the periodic table that would later be explained by an increasingly refined electronic structure of the atom and the concept of isotopy. Later in the century, Ostwald and his disciples were enthusiastic that physical chemistry would create a true *general* chemistry.

Nineteenth-century chemists also created a cultural and social structure for practicing chemistry in the academic and industrial worlds. The teaching–research laboratory began in the early nineteenth century, becoming a "factory" for producing chemists in large numbers. Despite the unity they saw in their discipline, chemists found themselves creating many subdisciplines with their own societies and journals. By the end of the century, chemistry was still predominantly male, but the new field of radioactivity, led by the example and role model of Marie Curie, inspired women from all over Europe and North America to enter chemical research laboratories around 1900. Although research in chemistry was still difficult for women, they were welcomed by the Curies in Paris, Rutherford at McGill University and Manchester, and, especially, by Stefan Meyer at the Vienna Radiation Institute (Malley 2011: 183).

The large number of new chemists would find careers in the burgeoning chemical industry, which by 1914 had been transformed in numerous ways. The production of inexpensive steel and aluminum made possible the growth of transportation networks, industrial machinery, and consumer goods. The new dye industry, spurred by the success of the structure theory, inaugurated the modern chemical industry and the science-based research and development laboratory, simultaneously creating and meeting the demand for fine chemicals and products for consumers. The growth and competition in the dye industry resulted from the implementation of robust patent laws and fierce competition between chemical firms in Germany. The emergence of the fine chemicals industry, together with the emerging electrical industry of the late nineteenth century (based on the newly developed electromagnetic theory), would lead to the material wealth of the modern world.

The chemical industry also created elaborate supply chains, from sulfuric acid to soda production to bleach, glass, and other materials, or the supply of coal tar for producing fine chemicals and dyes (and later polymers). Industrialists also realized that "waste" products, like hydrochloric acid from the Leblanc process or coal tar, could be used for other processes or turned into useful products, although there were also incentives in the form of newly created government regulations on the emission of toxic chemicals such as hydrochloric acid.

All three of these cultural trends – theoretical, social, and industrial – have clear turning points between 1850 and 1870, when the features of structural chemistry, production of alkali, dyes, and metals, and the means of training chemists, all started to coalesce into their modern form. By 1914, all the components and trends that would characterize chemistry in the modern age were in place. Chemists of the nineteenth century had created the culture and methods for mass production of many things – of chemicals, of chemists, of ideas, of subdisciplines – that would be carried into the twentieth century, when this mass production would accelerate exponentially as the demand for industrial chemicals, consumer goods, chemists, and specializations continued to increase (Agar 2012: 7).

CHAPTER ONE

Theory and Concepts: *Atomism, Structure, and Affinity*

TREVOR LEVERE

PRELUDE

By 1815, several developments provided a springboard for nineteenth-century chemistry. First, Antoine-Laurent Lavoisier had proposed an operational definition of chemical elements as substances that no one has yet been able to decompose by chemical analysis. Alongside his definition of elements was a new nomenclature of chemistry, in which the names of compounds reflected their composition, as in copper sulfate, where sulfate indicated a combination of sulfur and oxygen. Lavoisier also systematically used quantification as a regulative principle, insisting that the combined weights of the products of a chemical reaction had to equal the combined weights of the reactants, because matter was neither created nor destroyed in chemical processes. Second, between 1800 and 1810, the Voltaic pile, an early battery, became a standard instrument for chemical analysis. At the Royal Institution in London, Humphry Davy saw it as a tool of analysis that could decompose some of Lavoisier's supposed elements, and perhaps arrive at a smaller number of fundamental constituent elements. The chemical action of the battery also suggested that electricity and chemical affinity (the force that held the elements together in molecules or allowed chemical reactions to occur) were one and the same power.

The third theoretical innovation came from John Dalton, who suggested that for each of Lavoisier's elements, there existed a hard, unsplittable atom of a characteristic weight. Dalton observed that the combining weights of different elements for a given substance were in constant ratios, and that where several combinations of the same substances occurred their ratios could be arranged in a simple series. For example, carbon combined with oxygen to form two oxides, one containing twice the amount of oxygen as the other. He invoked simplicity criteria to argue that one had twice as many oxygen atoms as the other, and that because both oxides had the same weight of carbon, they should each have one atom of carbon. He thus invented a system of atomic weights and molecular formulas.

In 1809 in Paris, Joseph-Louis Gay-Lussac was led to a rival set of atomic weights by assuming that in equal volumes of gases there were equal numbers of particles. Water, for example, was formed from one volume of oxygen and two volumes of hydrogen. Gay-Lussac's volumes, however, led to a different set of atomic weights from Dalton's, and without a consistent set of atomic weights, it was impossible for chemists to agree on a single set of chemical formulas. A possible reconcilation came from Amedeo Avogadro in Italy, who in 1811 suggested that each particle of gases like hydrogen or oxygen consisted of two atoms, and that these particles split in two in chemical reactions. Avogadro's hypothesis provided a bridge between Gay-Lussac's law and Dalton's atomic theory and law of multiple proportions, but it would not be widely accepted until the 1860s.

Four major themes within nineteenth-century chemistry would flow from these basic ideas about elements, atoms, and affinity. First, by the 1860s, chemists would expand on the atomic theory and most of them would agree on a unified set of atomic weights. Second, driven by developments in organic chemistry, chemists would create sophisticated theories of the arrangement of atoms in molecules and arrive at formulas embodying structure, followed by an understanding of the three-dimensional arrangement of atoms in molecules (called stereochemistry). Third, the concept of chemical affinity would be reformulated as chemical thermodynamics in the attempt to create a new science of general or physical chemistry. Fourth, chemists were convinced that all areas of chemistry, organic and inorganic alike, should follow the same laws, and the newly discovered elements could be arranged systematically in the periodic table as developed by Dmitri Mendeleev. At the turn of the twentieth century, the discovery of radioactivity would eventually provide for a deeper understanding of atoms, atomic weights, and the periodic table. None of the themes above can be fully separated, because they often informed one another, but let us begin with atomic theory and atomic weights.

ATOMS, EQUIVALENTS, FORMULAS

Not everyone accepted Dalton's underlying assumptions of molecular simplicity, or his concept of interatomic attractions and repulsions. Whereas there was no problem in listing the relative combining weights of elements in molecules, or in recognizing the existence of multiple proportions (as in carbon monoxide and carbon dioxide), there were disagreements about the accuracy of Dalton's atomic weights, about the existence of diatomic molecules, and about the differences in atomic weights derived from Dalton's simplicity criteria and those derived from Avogadro's hypothesis.

There were about thirty known elements when Dalton published his theory. Hydrogen has the lowest atomic weight of all, so if, like Dalton, we arbitrarily assign an atomic weight of one to hydrogen, we can determine the weights of the atoms of other elements relative to that hydrogen standard. Dalton's dismissal of the possibility of diatomic molecules of any element meant that he regarded our diatomic molecule H_2 as an atom, not a molecule, and so his atomic weights of several other elements differed from ours by a factor of two. Oxygen had atomic weight seven for Dalton (more accurate experiments within Dalton's framework would have given it an atomic weight of eight); ammonia, our NH_3, being the lightest and therefore simplest combination of nitrogen and hydrogen, was Dalton's NH. The lightest and therefore simplest combination of carbon and hydrogen, methane, was CH_2 for Dalton, but CH_4 for us. Dalton's table of relative atomic and molecular weights thus involved inconsistencies and errors by modern standards. Some of the errors arose from Dalton's simplicity criteria. There were also errors arising from inaccurate experiments, especially those based on the gravimetric analysis of gases, for working with small weights of gases had always been difficult and prone to error for all but the best experimentalists. Another source of error might lie in the use of impure or contaminated substances.

Dalton's atomic and molecular weights were all listed as integers. Many atomic weights, when calculated on the arbitrary assumption that hydrogen has atomic weight one, are indeed nearly integers. Thomas Thomson and William Prout both believed that all atomic weights were integers. In 1815 and 1816 Prout published two papers in which he concluded that, if one took the atomic weight of hydrogen as 1, then the atomic weights of all other elements would be integral multiples of 1, and suggested that this was because the atoms of other elements might be formed of multiples of the hydrogen atom. He even conjectured that hydrogen itself might be compounded of smaller units. Thomson also published lists of integral atomic weights; in successive editions of his textbook *System of Chemistry*, he also supported Prout's hypothesis (Brock 1969; Brock 1985).

As chemists would determine nearly a hundred years later, there was a germ of truth in Prout's ideas, but the fact remained that there were some atomic weights that were far from integral. Thomson and Prout, for example, had given chlorine an atomic weight of 36, but the Swedish chemist Jacob Berzelius would calculate it to be 35.41, a result that remained roughly the same throughout the nineteenth century, making Prout's hypothesis difficult to accept. There is no reason to accuse Thomson of fraud in arriving at integral atomic weights. There is, however, a danger in knowing in advance of experiment what the result ought to be. From a series of experiments, one could select the result that best matched expectations; in Thomson's case, that could have meant choosing results that matched Prout's hypothesis that atomic weights would be integers. Those results could then be given to a high degree of precision, but that did not make them more accurate (Levere 2010).

One chemist who saw the perils of making assumptions about atoms and atomic weights, and who devised a way around those perils, was William Hyde Wollaston, who did know the difference between precision and accuracy (Rocke 1978; Rocke 1984). Wollaston accepted Dalton's laws of constant composition for compounds, and of multiple proportions when considering a series of different combinations of the same elements, but instead of using atomic weights, he used equivalent weights, which were a measure of combining ratios by weight. By using equivalent weights, Wollaston could avoid saying anything about Dalton's atoms. Equally, Wollaston was untroubled by non-integral atomic weights, because equivalent weights did not need to be integers. Chemists adopting Wollaston's equivalent weights were now free to determine equivalent weights as accurately as experimental methods allowed, without having to bring them in line with a speculative theory (Usselman 2015).

Chemical formulas, which for Dalton expressed the actual number of the different atoms in a molecule, could now be written to express the number of equivalent weights of different elements in a molecule; and analogous formulas indicated the existence of analogous properties, even if the actual number of atoms was unknown (Roberts 1976a). Sodium chloride and potassium chloride both contained one equivalent of metal and one equivalent of chlorine. Carbon dioxide contained one equivalent of carbon and two of oxygen, just as sulfur dioxide contained one equivalent of sulfur and two of oxygen. In 1814, Wollaston published a paper in which he described a logometric slide rule of chemical equivalents (Figure 1.1; Williams 1992). Such slide rules were widely used, and remained useful long after 1814. Michael Faraday, in his *Chemical Manipulation* (1827), promoted the instrument as being of great assistance to the laboratory chemist.

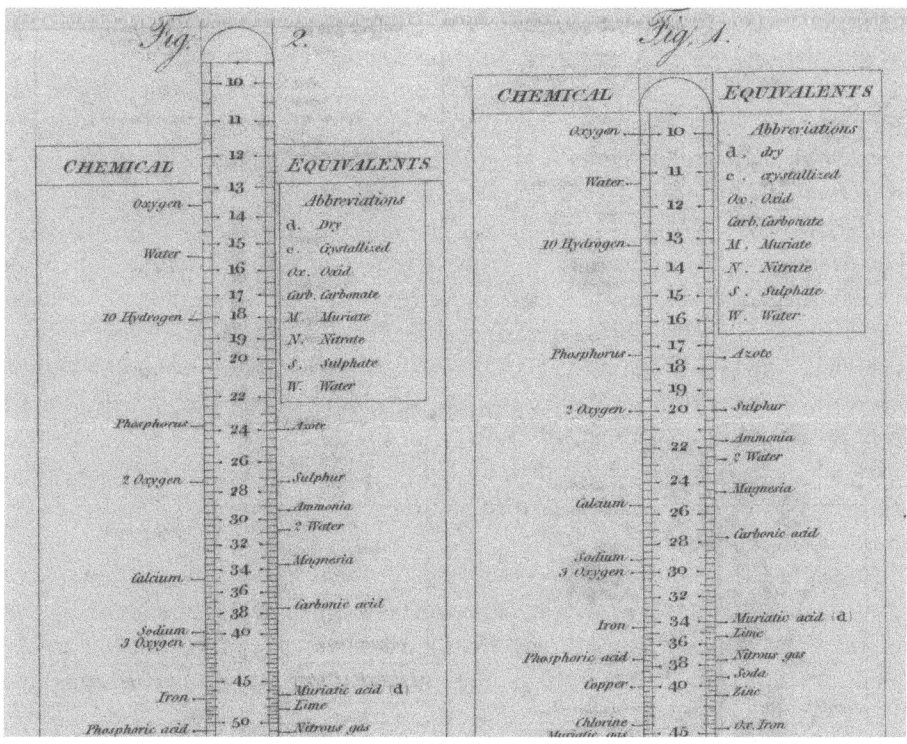

FIGURE 1.1 Wollaston's slide rule of chemical equivalents (Wollaston 1814). Courtesy of the Science History Institute, Philadelphia.

ELECTROCHEMICAL DUALISM

Davy used his electric battery to decompose compounds and to discover new elements, and showed that different chemical substances had different electrical properties. He regarded chemical affinity, the power of elements to combine, as the result of different degrees of electrical attraction, and argued that chemical affinity and electrical power might well be different manifestations of the same underlying force.

Davy was not the only chemist to explore the chemical action of electricity. He described the phenomena of electrolysis as "an alarm bell to the slumbering energies of experimenters in every part of Europe" (Davy in Levere 1971: 40). In Sweden, for example, Berzelius believed that the same laws operated on matter in both the organic and inorganic realms, and applied himself energetically to electrochemistry; he soon reached the conclusion that every element was distinguished by a unique electrical character. More than any other chemist, Berzelius brought together Lavoisier's conception of elements as the last products of analysis with electrochemistry and atomic theory. He arranged the elements in an electrochemical scale, without worrying whether they were ultimate elements or merely provisional elements; what mattered was their electropositivity or electronegativity, defined by the pole of the battery towards which they migrated. He believed that oxygen was the most electronegative element. Acid radicals, e.g. the sulfate in calcium sulfate or the nitrate in calcium nitrate, also had their own specific electrical character. This system allowed him to explain many chemical reactions. When sulfuric acid, with the highly negative sulfate radical, was added to an aqueous solution of sodium chloride, with its less-negative chloride radical, sodium sulfate was formed, along with hydrochloric acid. Berzelius' theory worked best for mineral salts, which had been central to much of eighteenth-century chemistry, and was particularly well suited to communities in Sweden and Germany where mining was a major industry.

Berzelius also introduced a new chemical notation by using letters for elements and superscript numbers to indicate the number of equivalents or atoms. He was by no means the first to adopt a systematic notation for chemical elements and compounds, but his notation was quickly adopted throughout Europe, spread through his textbooks and publications, and became a powerful tool for chemical discovery in the nineteenth century.

Berzelius initially organized mineral substances according to the electropositive nature of their constituent metals, which he argued determined their crystalline structure. Then, in 1819, the young Eilhard Mitscherlich in Berlin discovered isomorphism, in which one metal could replace another in a salt without substantially changing its crystalline form. It was as if the location of a metal atom in a salt was more important than its electrical character. Berzelius was both disturbed and impressed; he invited Mitscherlich to work with him in Stockholm. Convinced by Mitscherlich's experiments that minerals could no longer be classified by the electrical character of their constituent metals, Berzelius revised his classification of minerals so that the electrochemical nature of the complex acid radicals (e.g. sulfate, carbonate) became the key in determining the crystalline form and chemical reactions of minerals. He thus preserved electrochemical dualism as the explanation of the reactivity and properties of minerals (Levere 1971: 153–5).

Berzelius and Davy were the greatest chemists of the early nineteenth century, and both did important work in electrochemistry, but in many ways, it was Davy's one-time apprentice and successor Michael Faraday who would make the greatest advances in that branch of chemistry. Faraday, like his mentor Davy but unlike Berzelius, came to see himself as a natural philosopher, neither simply a chemist nor even a "scientist," a novel word coined in 1834 by the polymath

William Whewell. In Faraday's three volumes of *Experimental Researches in Electricity* (1839–1855), he argued that an electric current was a succession of arrangements and rearrangements of molecular forces within a conductor. He went on to develop an explanation of electrolytic transfer, and arrived at the conviction that, probably for all cases of electrochemical decomposition, "the chemical power of a current of electricity is in direct proportion to the absolute quantity of electricity which passes" (Faraday in Levere 1971: 83–7). He went on to define electrochemical equivalents as the proportions in which substances were evolved in electrolysis, which were the same as "ordinary chemical equivalents." Like Davy, he saw electrical attraction and chemical affinity as different manifestations of the same force, and, like Berzelius, he recognized the utility of arranging elements and compounds in an electrochemical scale.

CHEMICAL FORMULAS AND ATOMIC ARRANGEMENT

Isomorphism suggested that the chemical properties and even the composition of isomorphic crystalline substances might be correlated with crystalline form, although without revealing the actual arrangement of atoms in molecules (Levere 1975). If one accepted an atomic theory, then it was clear that atoms were somehow arranged and combined in molecules, and in some substances, molecules were assembled into crystals. There were competing versions of atomic theory: Dalton's atoms were indivisible, whereas Berzelius' atoms might be divisible; Dalton refused to contemplate diatomic molecules of an elementary gas, but Avogadro's hypothesis required divisible molecules. There were also competing versions of crystallographic theory: the geologist and mineralogist Abraham Gottlob Werner argued that a knowledge of the external characters of minerals yielded information about their composition, and the crystallographer René Just Haüy had produced a theory of crystalline structure that was contradicted by isomorphism. Even if one didn't believe in atoms, it was obvious that some substances existed in different and distinctive crystalline forms, but how could one begin to know what the arrangement of atoms really was in any substance? In other words, how could one create rational formulas indicating the arrangement of atoms?

Besides isomorphism, two other discoveries, isomerism and allotropy, became important in answering these questions about arrangement, although that was not immediately obvious. Isomers are compound substances with the same elemental composition, but with different properties and reactions. In 1826, Justus Liebig had prepared silver fulminate. In the following year, Friedrich Wöhler prepared silver cyanate, and found that it had the same chemical composition as the fulminate, but different properties. In 1830, Berzelius gave the name isomers ("same parts") to these compounds that had identical chemical formulas but different properties, which implied different arrangements of atoms, although no one knew what those arrangements might be. Some, like Berzelius, used reactions undergone by a substance as a way of arriving at formulas indicating possible linkages between atoms. His electrochemical dualism suggested that two strongly electronegative atoms (or, equally, two electropositive atoms) would repel one another, whereas two atoms of opposed electrochemical character (say hydrogen and chlorine, or sodium and oxygen) might be expected to join together. However, although chemists could agree about the composition of particular substances, there was no general agreement about formulas (Levere 1971: 169–75). A related problem was posed by the phenomenon of allotropy, where elemental substances exist in different physical forms. Charcoal and diamond, for example, were known to be pure carbon, but their physical properties and chemical reactivity were entirely distinct. Allotropy is yet another term invented by Berzelius, in 1841, when he wanted to describe the differences between red and yellow forms of mercuric iodide and the different physical forms of elemental sulfur (Jensen 2006).

ORGANIC CHEMISTRY

Organic chemistry is the chemistry of compounds containing carbon, combined with hydrogen, oxygen, and sometimes nitrogen. It began as the chemistry of living or formerly living matter, organic substances as opposed to minerals or metals, which had never been alive. The latter realm was incorporated into what became known as inorganic chemistry. The terms organic and inorganic came into use around 1820, and by then included not only minerals but also most gases. Chemists, aided by atomic theory, developed methods for determining the composition of organic substances, a necessary prelude to arriving at organic formulas.

There were good reasons why inorganic chemistry was more systematic and more advanced than organic chemistry in the early nineteenth century. First, pure inorganic substances are readily available: naturally occurring mineral ores may be a source of pure metals, salts, or oxides. Some elements, including sulfur and several metals, can be found in a state of high purity in nature, and these elements are part of the inorganic realm. Pure hydrogen may be generated by the dissolution of some metals in acids. Many of the elements involved in mineral chemistry are relatively heavy, and the number of atoms in an inorganic molecule is seldom very large, so that a small percentage error in analysis would not be likely to produce an error in the compositional formula of a mineral.

Contrast this with the circumstance that plant and animal substances are not found in pure form or in isolation: flesh, fat, leaves, manure, skin, and bodily fluids are complex, and it is not easy to extract pure substances from them. Solvent extraction, fractional crystallization, and distillation had been used for a century or more to try to isolate pure organic substances, and Lavoisier, among others, had also examined the products of combustion, with uneven success. Furthermore, organic molecules often contain very large numbers of atoms, and hydrogen, the lightest element, is usually the most numerous. That means that a small percentage error in analysis could result in compositional formulas with the wrong number of hydrogen atoms, and that in turn would lead to errors in identifying and classifying molecules.

In order to arrive at accurate compositional formulas for organic compounds, new analytical techniques were needed. Since Lavoisier, the determination of carbon and hydrogen had been carried out by combustion analysis, i.e. burning a carefully weighed sample of the substance under investigation, and then capturing and weighing the carbon dioxide and water produced. Knowing the atomic weights of carbon, hydrogen, and oxygen allowed the calculation of the amount of carbon and hydrogen in the sample. Two French chemists, Louis-Jacques Thenard and Joseph-Louis Gay-Lussac, developed a method based on volumetric analysis of the gaseous products of the combustion of organic substances.

In 1830, Liebig made the process of combustion analysis entirely gravimetric, avoiding the errors involved in volumetric analysis; a key advantage of gravimetric as opposed to volumetric analysis is that one could use much larger samples, which meant smaller errors. Liebig heated his sample, and collected the carbon dioxide produced by passing it through his newly invented potash apparatus (*Kaliapparat*; see Chapters 2 and 3 in this volume). He also developed what he considered to be an improvement on Dumas' method of estimating nitrogen, but this remained a volumetric method (Usselman et al. 2005). Still, Liebig arrived at consistent results for the composition of alkaloids and other substances, and proposed correspondingly accurate compositional formulas for them. Liebig was helped by his pedagogical innovation, the teaching and research laboratory, in which students could be trained in the techniques of analysis, and then be given different substances to analyze.

Along with the ease of the new methods of analysis, a variety of conceptual tools, particularly analogy, would be important in the development of theoretical organic chemistry. As early as 1814, Berzelius published the results of the quantitative analyses of some relatively simple organic substances, and it was clear to him that organic chemistry was underdeveloped. He believed that there was a single set of laws that would govern all of chemistry, organic and inorganic; that this chemistry was subject to the constraints of electrochemical dualism; and that the safe way to proceed was from the known to the unknown, from inorganic to organic chemistry. A fundamental assumption in Berzelius' inorganic chemistry was that the properties of a compound depended on the nature of the atoms of its constituent elements. If organic and inorganic compounds were analogous, then there must be groups of atoms that persevered through chemical reactions that could be considered, at least in some reactions, as if they were atoms. The acid radicals (sulfate, nitrate, etc.) of inorganic chemistry suggested by analogy that there might be similar complex organic radicals. In 1832, two German chemists, Justus Liebig and Friedrich Wöhler, announced their discovery of the benzoyl radical, which persevered through a variety of organic compounds, including benzaldehyde and benzoic acid.

This was strong support for what became known as the radical theory, but trouble was in store. In the 1830s, the French chemist Jean Baptiste André Dumas demonstrated that in several reactions, organic substances containing hydrogen, when treated with chlorine, lost hydrogen and, atom for atom, gained chlorine, without undergoing any major change in properties. To explain this, Dumas proposed a "theory of substitution": electropositive hydrogen could be replaced by electronegative chlorine. For example, acetic acid could be converted to chloroacetic acid or even trichloroacetic acid when chlorine was passed through it. Electronegative chlorine had replaced electropositive hydrogen, and yet the properties of the compound were almost unchanged, undermining the main assumption of the radical theory.

The substitution theory won energetic support from younger French chemists, who attacked electrochemical dualism and the radical theory head on. In 1837, Auguste Laurent wrote a doctoral dissertation proposing a new approach to organic chemistry, and he subsequently developed his argument in his posthumously published *Méthode de chimie* (1854). He spent his early career in Bordeaux, and had both the energy and the grievances that youth and provincialism implied. In matters of chemical theory, he was decidedly ambitious. He wanted the formula of an organic compound to represent the totality of its reactions, something that Berzelius' formulas could not do. A major problem was presented by substitution reactions, which Laurent made central to his argument. In 1844 he wrote a letter to Berzelius, arguing that the senior chemist had it all backwards (Brooke 1973). Laurent interpreted isomorphism (a discovery in inorganic chemistry) and isomerism (a discovery in organic chemistry) to mean that sometimes the position of atoms within a molecule was more important than their chemical nature. Arrangement, for Laurent, was the key to understanding organic chemistry; like Berzelius, he believed that there was only one chemistry, but in opposition to Berzelius, he argued that analogies based on organic chemistry should be applied to the whole of chemistry – a complete reversal of Berzelius' approach.

Laurent made common cause with Charles Frédéric Gerhardt, who also thought that substitution reactions were the key to determine the arrangement of atoms in a molecule. Similar formulas would then represent similar arrangements and lead to similar reactions, and molecules with similar arrangements could be arranged in series. However, Gerhardt went further, and argued in 1842 that in order to make formulas consistent throughout organic chemistry, it was necessary to represent water not as H_4O_2, but as H_2O. This meant that many formulas for organic compounds would have half the number of atoms. Gerhardt also redefined acids as substances with displaceable hydrogen, thus accounting for the formation of salts; in the year of Laurent's death, he published his theory of "types" (Figure 0.1): hydrogen, H−H, with two atoms; water, H_2O; and ammonia, NH_3. Ethyl alcohol (C_2H_5OH) could be seen as derived from the water type by substitution of an ethyl radical (C_2H_5) for one of the hydrogen atoms in water. Thus types, which Gerhardt had proposed as a guide to organic formulas and reactions, were based on analogy to simple inorganic molecules, and provided a means of unifying organic and inorganic chemistry.

Berzelius and Liebig thought electrochemical dualism and the concept of radicals would be the path into the new and difficult realm of organic chemistry that would create a unified chemistry. Laurent, on the other hand, thought in terms of a structural dualism, in which molecules contained an inner part that could not undergo chemical change without altering the properties of the substance, and an outer part that could undergo chemical change while preserving the overall properties of the substance. We have seen that the substitution of chlorine for hydrogen in a methyl radical in acetic acid still left the molecule as an organic acid; this was not a problem for Laurent and Gerhardt, but it did at first pose a problem for Berzelius and Liebig.

Gradually, types and radicals became indistinguishable in chemical theory. In 1853, the year of Laurent's premature death, Liebig wrote to Gerhardt that "it is very strange that the two theories, formerly quite opposed, are now combined in one which explains all the phenomena in the two senses" (Liebig in Partington 1961–1970: vol. 4, 460), but this unification could only have occurred by the substantial modification of both theories. The properties of chemical compounds were determined both by the nature of their constituent atoms and by the arrangement of atoms within molecules.

Edward Frankland, for example, combined formulas based on types with electrochemical ideas about atoms (Russell 1996). He used type formulas for organic and inorganic substances alike, so that phosphorus trihydride and tri-ethyl phosphorus, belonging to the same type, had analogous formulas. He noted that there was a clear analogy between many stable compounds, e.g. the trihydrides of nitrogen, phosphorus, and arsenic. This led him to theorize that "no matter what the character of the uniting atoms may be, the combining power of the attracting element, if I may be allowed the term, is always satisfied by the same number of these atoms." Frankland's contemporaries called this "combining power" atomicity, and it is perhaps the earliest reference to the concept of valence. This insight into the combining powers of different elements involved a new way of thinking; previous chemists had been concerned with different combinations of the same elements, e.g. the oxides of nitrogen, rather than with parallels between formally similar series relating to different sets of compounds. Frankland's belief in the unity of organic and inorganic chemistry was reinforced by his discoveries in the new field of organometallic chemistry, where the electrochemical nature of constituent atoms and the existence of types were decisive in arriving at the proper classification of compounds (Levere 1971: 189–91; Russell 1971).

AGREEMENT ON ATOMIC WEIGHTS

Creating reliable and accurate rational formulas, for organic and inorganic compounds alike, relied on accurate compositional formulas, theories of atomicity and types, and above all, the atomic weights of constituent elements. By mid-century, atomic weights had become more precise, and by 1860, the German-trained Swiss chemist Jean Charles Galissard de Marignac had published results of atomic weights to three decimal places. Chemists, however, still had to agree on a single set of atomic weights. This issue was not a question of arguing about decimal points. Which set of weights should be used as a universal standard? Should given sets of atomic weights be left alone, halved, or doubled? One lead had been offered by Pierre-Louis Dulong and Alexis-Thérèse Petit, who in 1819 published a paper showing that, in the case of twelve metals and sulfur, the product of atomic weight and specific heat was a constant value. To arrive at this result, which they generalized as the law of atomic heats, they had to change several of Berzelius' atomic weights, for example halving his atomic weights for sulfur and silver (Partington 1961: vol. 4, 200–1).

The eventual key to consistent atomic weights would be to coordinate the results of the Dulong–Petit law with Avogadro's hypothesis, which was first proposed in 1811, but received little attention (Brooke 1981). Dumas had given support to Avogadro's hypothesis in 1826, and Marc Antoine Augustin Gaudin also recognized its importance for atomic weights in a paper sent to the Académie des Sciences in 1833. His work was all but ignored until 1858, when the Italian chemist Stanislao Cannizzaro, fully recognizing Gaudin, showed that Avogadro's hypothesis could lead to a rigorous and coherent system of atomic weights.

Cannizzaro's paper also had little impact until 1860, at the first international conference of chemists, held in Karlsruhe, Germany, organized by August Kekulé, Adolphe Wurtz, and Karl Weltzien. On the last day of the conference, Cannizzaro presented his argument, and his friend Angelo Pavesi distributed copies of his 1858 pamphlet. Some chemists were immediately convinced, some missed the point, and others read the pamphlet only later. The German chemist Lothar Meyer was among the last group. He read Cannizzaro's pamphlet on the train on his journey home. "I was astonished," he wrote, "by the clearness that the brochure provided concerning the most important points at issue. It was like blinders being removed from from my eyes; doubts disappeared, and a quiet feeling of certainty replaced them" (Rocke 1993: 203). The Karlsruhe conference and the eventual certainty of most chemists that Canizzaro was correct were turning points for accepting Avogadro's hypothesis. Chemists finally agreed on a clear distinction between atoms and molecules, such that the gaseous elements nitrogen, oxygen, and hydrogen all consisted of diatomic molecules. Adopting Avogadro's hypothesis also allowed them to agree finally on a single set of atomic weights for all the elements, in which, for example, oxygen was sixteen, not eight.

ORGANIC FORMULAS REFINED

All of these factors – acceptance of Avogadro's hypothesis, the merging of radical and type theory, and the concept of valency – would come together by the early 1860s in the form of the structural theory of organic chemistry, in which chemical formulas would indicate specific linkages or bonds between atoms within molecules. The idea of interatomic linkages in molecules was itself controversial. Gerhardt, who had attended Auguste Comte's lectures, thought that one should not speculate about the ways in which atoms were combined. However, not all chemists were averse to speculation. Alexander Williamson, an English chemist of Scottish descent – ironically, also a student of Comte – wrote as early as 1852: "Formulas ... may be used as an actual image of what we rationally suppose to be the arrangement of constituent atoms in a compound, as an orrery is an image of what we conclude to be the arrangement of our planetary system" (Williamson in Russell 1971: 51).

A key to the concept of structure would be the introduction of the carbon type and the rise of the theory of chemical structure in 1858. The carbon type appeared independently twice in 1858, by Kekulé, and by Archibald Scott Couper (Benfey 1963). Kekulé had already asserted the quadrivalence of carbon in 1857, adding a fifth type to Gerhardt's four: the new type represented the combination of four atoms or radicals with one atom of carbon. In his 1858 paper he stressed that one needed to consider the valence of all the individual atoms in arriving at molecular formulas. Couper's paper represented the valency bonds between atoms by dotted lines, and stated that an atom of carbon combined with equal numbers of equivalents of hydrogen, oxygen, etc. In a crucial second major innovation, he stated, as Kekulé had also explained in his concurrent paper, that carbon atoms could use their valence units to combine with other carbon atoms, so that organic molecules could be structured around a molecular chain or backbone of carbon atoms. This was the essential reason why there were so many organic compounds, and why most of them were so complex. Couper and Kekulé had independently and almost simultaneously proposed the immensely consequential theory of chemical structure, but Couper's paper appeared slightly later than Kekulé's, and it was the latter whose paper was more influential (Rocke 2010).

In the 1860s, chemists introduced a variety of graphic representations of the linkage of atoms within molecules. Kekulé himself used what have been called "sausage" formulas in his textbook, although he did not use them often (Figure 1.2). The most successful early representation of structure was introduced by Scottish chemist Alexander Crum Brown, using simple lines to note bonds between atoms. Using this notation, the two-carbon compounds ethane ethylene and acetylene could easily be depicted with single, double, and triple bonds between carbon atoms (Figure 1.3). One of the greatest early successes of structural formulas was Kekulé's resolution of the structure of benzene; in 1865, he proposed a ring structure for it, initially with a modified version of his sausage formulas, and later using Crum Brown's formulas showing alternating single and double bonds (Figure 1.4). The cyclic structure for benzene and its derivatives proved to be a remarkably effective tool for predicting the existence of new

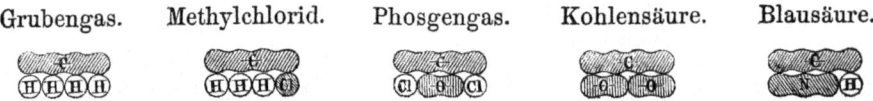

FIGURE 1.2 Kekulé's "sausage" formulas. Bonds between atoms are shown by those points of atoms which touch vertically (Kekulé 1861: 162). Courtesy of the Oesper Collection, University of Cincinatti.

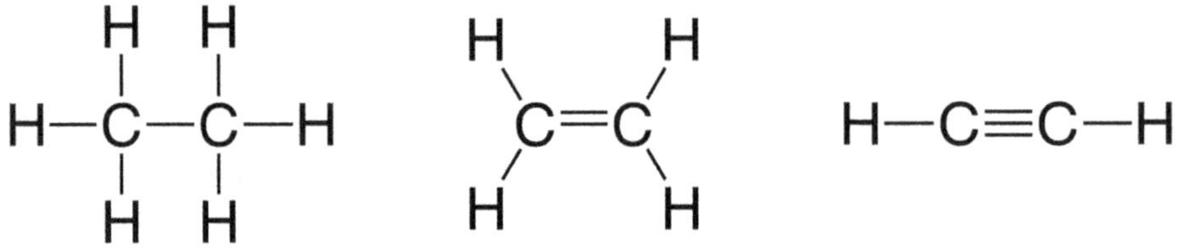

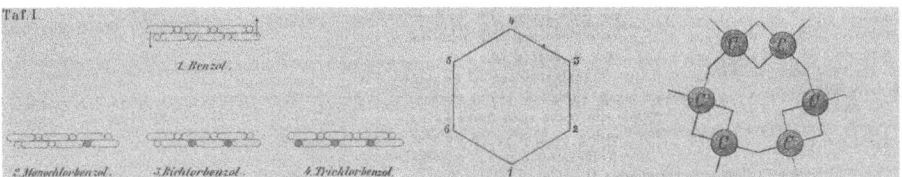

FIGURE 1.3 Formulas for ethane, ethylene, and acetylene using Crum Brown's symbolism. Drawing by Peter J. Ramberg.

FIGURE 1.4 Kekulé's early cyclic formulas for benzene (Kekulé 1867: 4, 22, 6–7). Courtesy of the Science History Institute, Philadelphia.

compounds, ultimately leading to the foundation of the new synthetic dye and pharmaceutical industries of the late nineteenth century.

Until 1874, all of these graphic or structural formulas indicated the arrangements of the atoms in a molecule in two dimensions. In fact, chemists were very explicit that these formulas were not meant to be a direct image of the microworld of the molecule. In 1874, two papers were published, again quite independently, that suggested that formulas could indeed represent the three-dimensional arrangement of atoms in space. The authors were, respectively, the Dutch chemist Jacobus Henricus van 't Hoff and the French chemist Joseph Achille Le Bel. Although van 't Hoff's paper was first published in Dutch and remained obscure, later editions in French and German proved to be more influential among chemists than Le Bel's paper (Ramberg and Somsen 2001; Ramberg 2003). Van 't Hoff's argument was grounded in the traditional explanation of isomerism, in which the same atoms could be arranged differently in different substances, but van 't Hoff extended the meaning of "arrangement" to atoms in three-dimensional space. If methane, CH_4, underwent substitution of three of its hydrogen atoms by three different organic radicals ($CHR_1R_2R_3$), then planar formulas or models indicated that there should be three different isomers, because one could write three different distinct planar arrangements of these structures; but there were not three different isomers. Indeed, before 1874 chemists generally agreed that formulas and models (including some ball-and-stick models) were essentially just conventions, and did not represent the three-dimensional arrangement of atoms in space. Van 't Hoff and Le Bel both argued that a carbon atom really did have its four bonds directed to the corners of a tetrahedron, with the carbon atom at the center. In that case, the molecule $CHR_1R_2R_3$ would have two isomers, because there were two and only two three-dimensional arrangements of the atoms and radicals; the two arrangements were distinct, because they were asymmetrical and thus not superimposable (Figure 1.5). Both Van 't Hoff and Le Bel noted that the two asymmetric molecules would rotate the plane of polarized light in equal but opposite directions, a property that was well known but previously unexplained at the molecular level. Jean-Baptiste Biot had first observed the phenomenon of "optical activity" in 1815, and in 1848, Louis Pasteur had studied the optical activity of crystals of certain tartaric acid salts, and suggested that the cause of optical activity was asymmetric molecules. But it was only in 1874 that optical activity was explicitly linked to a specific structural feature of an organic molecule. Van 't Hoff and Le Bel's arguments were underpinned by the physical property of optical rotation.

Although there was some skepticism, and, in the case of Hermann Kolbe, outright hostility toward Van 't Hoff's theory, the overall reaction was one of cautious acceptance of a theory that showed enormous promise (Rocke 1993: 329; Ramberg 2003). In 1893, Alfred Werner expanded stereochemistry to

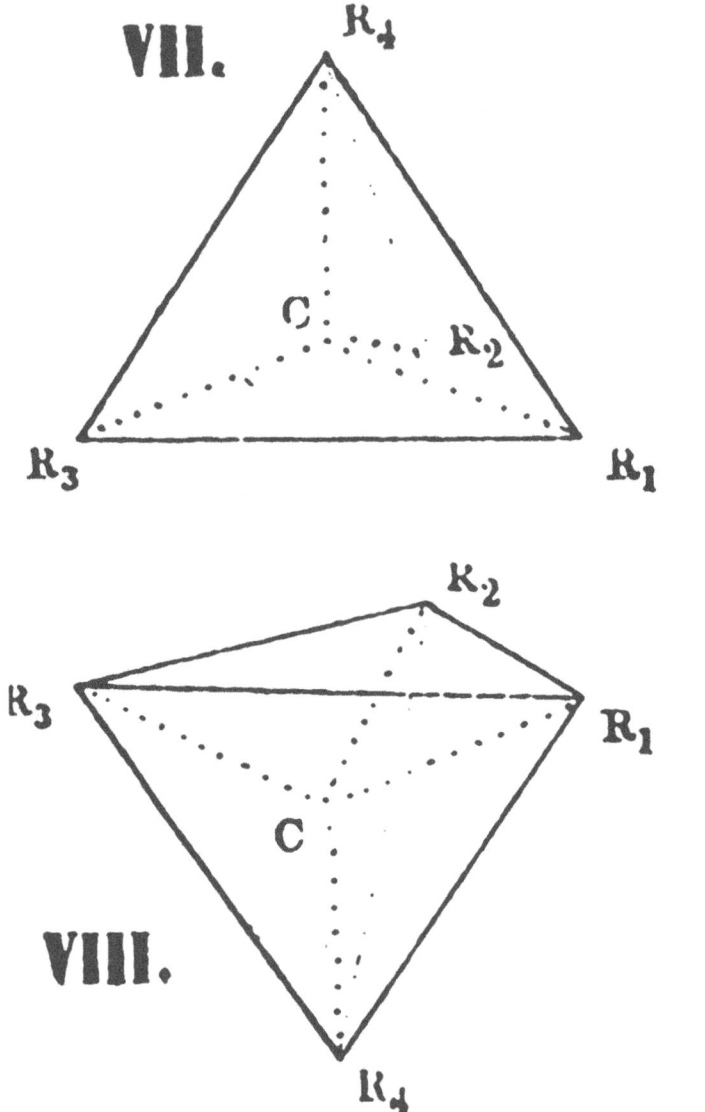

FIGURE 1.5 Van 't Hoff's depiction of the tetrahedral carbon atom. The top is a mirror image of the bottom, and the two tetrahedra are not superimposable (van 't Hoff 1874). Courtesy of the Oesper Collection, University of Cincinatti.

the inorganic coordination compounds containing complex ions (Kauffman 1966; Kauffman 1968).

Chemists had now agreed on the concept of the arrangement of atoms and had accepted a consistent set of atomic weights, but how could they establish the relations between atomic weights and the properties of atoms? How could one arrange chemical species in a way that brought out analogies and made sense? We need to step back, and consider approaches to these questions, in particular that of Mendeleev and his periodic law, embodied in the periodic table.

MENDELEEV AND THE PERIODIC TABLE

The agreement on atomic weights resulting from the Karlsruhe conference would also have a profound effect on the organization of the known elements, resulting in the famous periodic table created by Dimitrii Mendeleev in 1869. The concept of periodicity had long predated Mendeleev, however. In 1829, the German chemist Johann Wolfgang Döbereiner had already observed that there were several groups of three elements, triads, where the atomic weight of one of the elements was very nearly halfway between the atomic weights of the other two. These triads were of recognizably similar elements, including the halogens (chlorine, bromine, and iodine) and the alkaline earth metals (calcium, strontium, and barium). The lack of agreement about atomic weights made it impossible to extend Döbereiner's thinking to many more elements.

After the standardization of atomic weights, several chemists published similar kinds of regularities in the properties of elements. During the 1860s, chemists would number the elements and arrange them in tables, write them in a line wrapped around a cylinder, plot them on graphs, and create a dozen ingenious ways of displaying periodic regularities in atomic weights and bring out analogies between the properties of the elements. Lothar Meyer discovered that if elements were arranged according to their atomic weights, they emerged as groups in which analogous chemical and physical properties were repeated at periodic intervals (Spronsen 1969; Knight 1970; Levere 2001; Scerri 2007).

However, Mendeleev would introduce the most comprehensive and enduring contribution to the concept of periodicity. Mendeleev had attended the Karlsruhe Conference in 1860, where he met Cannizzaro, Wurtz, Dumas, and other leading European chemists. After his return to Russia, while writing a textbook of organic chemistry, he realized that organic compounds could be arranged in a table, with the homologous series in the horizontal lines and different classes of compounds (paraffins, alcohols, chlorides, etc.) vertically. In 1868, writing a textbook of general chemistry, he looked for a comparable way of classifying elements, and found it when he arranged groups of elements in the order of their atomic weights, e.g. for the alkaline earths, calcium (atomic weight 40), strontium (87.6), barium (137), and for the halogens fluorine (19), chlorine (35.5), and bromine (80). Mendeleev's use of analogy between organic compounds and inorganic elements reflects again the desire of chemists for unifying organic and inorganic chemistry.

Mendeleev first stated his periodic law in 1869: "Elements placed according to the value of their atomic weights present a clear periodicity of properties" (Mendeleev in Kedrov 1981: 288). Later, Mendeleev more fully described his law: "The properties of the elements, as well as the forms and properties of their compounds, are in periodic dependence or (expressing ourselves algebraically) form a periodic function of the atomic weights of the elements." Mendeleev then arranged these periods into a general pattern that would include all the elements, initially with rows for analogous elements with the same valency and columns for increasing valencies. This would become the periodic table of the elements (Kedrov 1981; Kaji 2003; Figure 1.6).

Mendeleev had the confidence to leave blank spaces in his periodic table, and to predict that those spaces would be filled by as yet unknown elements, whose properties he boldly specified. Some of his predictions were borne out by the discovery of new elements in laboratories across Europe. For example, there was a blank space after zinc, in the same group as boron and aluminum. Mendeleev predicted that this blank would be filled by an element with properties similar to aluminum and having an atomic weight of 68 and specific gravity of 6.0. In 1875, the missing element was discovered by Paul-Emile Lecoq de Boisbaudran in Wurtz's laboratory, who called it gallium, after the French nation. He also may have named it after himself, as "Lecoq" in French and "Gallus" in Latin both mean rooster, and the rooster is the national symbol of France. Gallium was found to have an atomic weight of 69.9 and a specific gravity of 5.96, almost exactly as Mendeleev had predicted. Scandium and germanium were two other previously unknown elements whose properties were predicted by Mendeleev with equal accuracy (Levere 2001: 119). There were still difficulties and problems to resolve, for example just what to do with the rare earths or lanthanides, whose atomic weights were not always reliable. There were also problems with some pairs of elements. Tellurium, for example, with an atomic weight of 128, should have come after iodine, with atomic weight 127; but that would have put it with the halogens, rather than with oxygen, sulfur, and selenium. Here, Mendeleev made periodicity and analogous properties the crucial factor, and so he overrode the atomic weights. Over the years, Mendeleev made successive changes and improvements to his table, revising it as he revised his textbook, and the table and the periodic law became more central to an understanding of the chemical elements and their properties (Kaji et al. 2015).

Late in the century, the periodic table enabled chemists to handle the discovery of the so-called inert gases (now called noble gases) with aplomb. In 1894, at the meeting in Oxford of the British Association for the Advancement of Science, Lord Rayleigh (Robert John Strutt) and William Ramsay announced the discovery of argon, which appeared to be completely unreactive. Ramsay had already written to Rayleigh: "Has it occurred to you that there is room for gaseous elements at the end of the first row of the periodic table?" (Weeks 1968: 784). Ramsay, with help from William Crookes, confirmed the discovery

— 70 —

но въ ней, мнѣ кажется, уже ясно выражается примѣнимость выставляемаго мною начала ко всей совокупности элементовъ, пай которыхъ извѣстенъ съ достовѣрностію. На этотъ разъ я и желалъ преимущественно найдти общую систему элементовъ. Вотъ этотъ опытъ:

```
                              Ti=50    Zr=90      ?=180.
                              V=51     Nb=94      Ta=182.
                              Cr=52    Mo=96      W=186.
                              Mn=55    Rh=104,4   Pt=197,4
                              Fe=56    Ru=104,4   Ir=198.
                           Ni=Co=59    Pl=106,6   Os=199.
         H=1                  Cu=63,4  Ag=108     Hg=200.
              Be=9,4  Mg=24   Zn=65,2  Cd=112
              B=11    Al=27,4 ?=68     Ur=116     Au=197?
              C=12    Si=28   ?=70     Sn=118
              N=14    P=31    As=75    Sb=122     Bi=210
              O=16    S=32    Se=79,4  Te=128?
              F=19    Cl=35,5 Br=80    I=127
    Li=7      Na=23   K=39    Rb=85,4  Cs=133     Tl=204
                      Ca=40   Sr=57,6  Ba=137     Pb=207.
                      ?=45    Ce=92
                      ?Er=56  La=94
                      ?Yt=60  Di=95
                      ?In=75,6 Th=118?
```

а потому приходится въ разныхъ рядахъ имѣть различное измѣненіе разностей, чего нѣтъ въ главныхъ числахъ предлагаемой таблицы. Или же придется предполагать при составленіи системы очень много недостающихъ членовъ. То и другое мало выгодно. Мнѣ кажется притомъ, наиболѣе естественнымъ составить кубическую систему (предлагаемая есть плоскостная), но и попытки для ея образованія не повели къ надлежащимъ результатамъ. Слѣдующія двѣ попытки могутъ показать то разнообразіе сопоставленій, какое возможно при допущеніи основнаго начала, высказаннаго въ этой статьѣ.

Li	Na	K	Cu	Rb	Ag	Cs	—	Tl
7	23	39	63,4	85,4	108	133		204
Be	Mg	Ca	Zn	Sr	Cd	Ba	—	Pb
B	Al	—	—	—	Ur	—	—	Bi?
C	Si	Ti	—	Zr	Sn			
N	P	V	As	Nb	Sb	—	Ta	—
O	S	—	Se	—	Te	—	W	—
F	Cl	—	Br	—	J	—	—	—
19	35,5	58	80	190	127	160	190	220.

FIGURE 1.6 Mendeleev's first Periodic Table of Elements, 1969. Photograph by Ann Ronan Pictures/Print Collector/Getty Images.

of helium in 1895. After hearing of the discovery of argon, de Boisbaudron also predicted that it might belong to a family of inert gases and predicted their atomic weights, with values very close to those found for the as yet unknown elements neon, krypton, and xenon, discovered by Ramsay in 1898 (Weeks 1968: 778–801). In light of Ramsay's discoveries in 1898, Mendeleev accepted the new elements, which now had their own column in the periodic table; indeed, he came to see Ramsay's work as strongly reinforcing the periodic law (Kedrov 1981).

PHYSICAL CHEMISTRY

Although issues of atomic weights and molecular arrangement dominated theoretical issues throughout the nineteenth century, a significant number of chemists asked different questions. One set of questions concerned the relationship between chemical and physical properties. Hermann Kopp, for example, showed how the boiling points of organic compounds increased steadily with the number of carbon atoms, and in 1864 he showed that the specific heats of compounds could be directly calculated from the specific heats of their elements. As we saw above, Biot, Pasteur, Le Bel, and van 't Hoff had established a relationship between chemical composition and optical activity caused by the spatial arrangement of atoms.

Another set of questions revolved around the attempt to understand how and why chemical reactions occurred. While it was clear that atoms were held together in molecules, it was unclear what forces held them together and what would break those forces during a chemical reaction. Early in the century, chemists had already been convinced that chemical affinity was electrical. In the 1830s, Michael Faraday established laws of electrochemistry. In the next generation, Friedrich Wilhelm Georg Kohlrausch, who held a succession of appointments in physics, carried out precise measurement of electrical conductivity during electrolysis.

Other chemists attempted to solve the problem of chemical affinity by measuring the heat generated by chemical reactions. Julius Thomsen made an exhaustive study of thermochemistry, eventually gathering the results in four volumes, published in 1882–1886. He suggested that the heat produced during chemical reactions was a measure of the chemical affinity between atoms in the reactants. Thomsen found himself engaged in priority disputes with Marcellin Berthelot, who had independently reached the same conclusion (Kragh 2016). Much of their work, like that of Davy and Faraday, can be seen as an exploration of chemical affinity, which would come to fruition in the development of chemical thermodynamics.

It is apparent that physical and theoretical chemistry, although not yet recognized as a new discipline, had wide and deep roots throughout the nineteenth century. It emerged as a discipline in the work of van 't Hoff, the Latvian-born German Wilhelm Ostwald, and the Swede Svante Arrhenius. Chemical thermodynamics and chemical kinetics, encompassing reactions in the gas phase and in solution, were at the core of the new physical chemistry (Nye 1993; Laidler 1995; Hirota 2016).

Ostwald, van 't Hoff, and Arrhenius all wanted to move away from taxonomy and the study of chemical substances (as we have seen, a major concern for both organic and inorganic chemistry), and toward exploring the principles that governed the formation of those substances and their reactions. They wanted to develop a general chemistry through an application of physical methods to chemical problems. Because of their work on electrolytic dissociation and Arrhenius' novel theory of ionic solutions, they became known collectively as "the ionists." They investigated the effects of mass and temperature on chemical equilibrium, osmotic pressure, and electrolytic conduction, and in the early 1880s, their work coalesced into a coherent program of research and theory (Servos 1990: 21).

Arrhenius wrote his dissertation on the electrolysis of solutions at the University of Uppsala, sent copies to Ostwald and Van 't Hoff, studied with the former in Riga and the latter in Amsterdam, and then returned to Sweden, where for the remainder of his career he taught in Stockholm. They remained in close touch and succeeded so well in their subdiscipline that all three would receive Nobel prizes in chemistry, with Van 't Hoff in 1901 as the first chemist to be so honored.

It should be noted that the ionists did not invent chemical thermodynamics. Josiah Willard Gibbs, professor of mathematical physics at Yale, may be considered its founder. He worked largely in isolation, partly because of the complicated notation that he used, which few knew about and fewer understood. Others, working independently, covered the same ground more accessibly. But it was van 't Hoff, whose approach was less rigorous than Gibbs', but more chemical and more practical, who wrote the first accessible monograph in the field, *Études de dynamique chimique* (1884), treating chemical thermodynamics and chemical kinetics. The book encompassed reversible reactions, dynamic chemical equilibrium, and equilibrium constants, and explored both gases and solutions. Chemical affinity was central to van 't Hoff's *Études*, which showed how temperature, pressure, and mass were related to rates of reaction, how heats of reaction could be calculated from the final state of equilibrium, and how equilibrium was dependent on temperature; all these terms were now defined in terms of the work that a reversible chemical reaction could perform, and yielded a new definition of an old concept, chemical affinity.

Ostwald also used thermodynamics as a way to avoid having to deal with atoms. In 1896, he observed: "We judge a process by the kind and quantity of the vanishing and appearing energies. These we are able to measure, and all that is necessary to know may be expressed in this manner" (Ostwald in Nye 1984: 352). By 1904 he had been able to show that all stoichiometric laws could be determined without invoking atomism, basing his argument instead on three pillars: Gibbs' phase rule governing the equilibrium between phases (derived from thermodynamics); van 't Hoff's law expressing the effect of temperature on chemical equilibrium; and the principle of Henri Louis Chatelier, which should really be called van 't Hoff's principle, showing how a shift in conditions would lead to a shift in chemical equilibrium (Laidler 1995: 117–18).

In 1814 Joseph Fraunhofer was the first to study and make an accurate record of the dark lines in the solar spectrum. In 1847 the English-born John William Draper, whose scientific career was wholly in America, published a paper stating principles of spectroscopy that were subsequently given mathematical form by the German physicist Gustav Kirchhoff at Heidelberg (Laidler 1993: 178). Kirchoff and the chemist Robert Bunsen collaborated in the development of chemical analysis using a spectroscope of their own design. They and other chemists soon used the instrument to identify known elements and predict new ones. The method was then applied to the analysis of stellar and solar light by the astronomers Joseph Norman Lockyer and William Huggins (McGucken 1969).

DIVISIBLE ATOMS, ATOMIC STRUCTURE, RADIOACTIVITY, AND ISOTOPES

By 1910, almost all chemists (including Ostwald) had come to accept the reality of atoms, but as the extensive study of cathode rays and the new phenomena of radioactivity had shown, atoms were not Dalton's simple, homogeneous, indivisible spheres. Working with cathode ray tubes in 1897, the English physicist Joseph J. Thomson found that there was a stream of negatively charged particles between the electrodes at each end of the tubes. Thomson measured the ratio of the electron's mass and electric charge, and named the particle the electron. Following Thomson's discovery, physicists gradually realized that there also had to be positively charged particles. In 1911, the New Zealand-born physicist Ernest Rutherford, then a professor in Manchester, showed that the bulk of an atom's mass was a positively charged nucleus that made up very little of the atom's volume (Hirota 2016: 193–8).

In 1895, the German physicist and engineer Wilhelm Conrad Röntgen had also been studying cathode rays and found that they caused the fluorescence of nearby crystals in his laboratory. He also found that photographic plates, although wrapped in opaque paper to protect them from light, were fogged. He realized that a new kind of ray was the cause of the fluorescence and the fogging, and called them x-rays because their nature was mysterious. In the following year, the French physicist Henri Becquerel found that uranium salts likewise emitted radiation capable of passing through opaque paper and affecting photographic plates; in doing so, he discovered radioactivity. In 1898, the Polish-born scientist Marie Skłodowska Curie and her French husband Pierre built on Becquerel's discovery of radioactivity, and in the course of their researches discovered radioactive radium and polonium, and coined the term "radioactivity" to describe this mysterious emanation (Weeks 1968; Emsley 2003).

Rutherford also studied radioactivity, beginning with radium and thorium. In 1899 he found that there were two different kinds of rays emitted by radium, one that could penetrate thick sheets of aluminum, and another that could only penetrate thin sheets. He called these β rays and α rays, respectively. Working with Frederick Soddy at McGill University in Montreal, by 1901 he was able to show that radioactivity was produced by the spontaneous decay or disintegration of the atoms of a radioactive element, and was accompanied by the formation of other elements. Soddy went on to work with Sir William Ramsay, and showed that the decay of radium produced helium, and later showed that uranium decays to form radium (Figure 0.5). In 1913, he coined the term "isotope" for variants of the same elements with different atomic weights.

This revision of atomism entailed by such discoveries would in turn open the way to a new understanding of the electronic structure of atoms, the periodic table, chemical affinity, chemical bonding, and molecular structure. After Thomson's work, chemists came to accept that electrons held the answer to the problem of chemical affinity. But what did electrons do? Chemists didn't have an answer, although they might have found a starting point in the work of the American physical chemist Gilbert Newton Lewis, who in 1902 sketched out an idea about bonding that remained unpublished until 1916. He represented atoms as cubes, with electrons at the corners, and drew cubes for lithium, beryllium, boron, carbon, nitrogen, oxygen, and fluorine, elements in the second period in the periodic table, with the same cubes for elements of the third period, magnesium, aluminum, silicon, phosphorus, and sulfur (Lewis 1923: 29). He started with one electron for lithium, adding an electron as he moved across the group, and ending with seven electrons for fluorine. The most stable state of combination was when electrons were shared between atoms in such a way that each atom had eight electrons at the corners of its cube. In a single bond, two atoms shared a single electron; in a double bond, two atoms shared two electrons (Levere 2001: 171–2).

In 1904 Richard Abegg, a Polish-born physical chemist who had no way of knowing about Lewis' unpublished sketch, produced a table showing a striking and simple correlation between the variable valencies of elements and the group in the periodic table to which those elements belonged. Elements had a normal or primary valence equal to the number of their group, and a secondary valence equal to the difference between the primary valence and eight. Lewis published his ideas on the role of electrons in chemical combination in 1916, and they can be seen to be consistent with Abegg's, but it was not until the development of quantum mechanics and wave mechanics that chemists had a mathematical way of tying electrons to valence bonds (Levere 2001: 173).

CONCLUSION

During the long nineteenth century, Daltonian atomism led to vigorous debates between the use of true atomic weights and more empirically oriented chemical equivalents, although both approaches were extremely productive in creating formulas. There were some chemists who refused to accept chemical atoms, for example, Michael Faraday and Wilhelm Ostwald. But in the end, the heuristic value of atomism outweighed any philosophical scruples. For atomism to be fully productive, an accurate, coherent, and self-consistent set of atomic weights was essential, but not until chemists agreed about the fundamental criteria for calculating them. This changed thanks to Cannizzaro's paper at the Karlsruhe conference in 1860, making clear the importance of Avogadro's hypothesis. Subsequently, the vast majority of chemists accepted chemical atoms, although a few rejected them, most prominently Wilhelm Ostwald. The last decade of the nineteenth century and the first decade of the twentieth century saw the discovery of the electron, of radioactivity, and of transmutation, indicating that atoms were divisible.

It was relatively easy for chemists to agree that molecular formulas based on atomic weights were possible, but it was harder for them to agree about specific molecular formulas, partly because of competing sets of atomic weights. Classification of compounds suggested analogies between the same kinds of substances in both inorganic and organic chemistry. The radical theory brought order to inorganic chemistry, and Berzelius, Liebig, and others tried to extend the concept of radicals into organic chemistry. On the other hand, theories of organic chemistry, developed by Dumas, Laurent, and Gerhardt, brought an alternative order via the type theory to large groups of organic compounds. In the 1850s, thanks to the work of Frankland and others, the two theories converged, making possible a truly unified chemistry. Analogies, however, were not enough. How were the atoms connected in molecules? What was the structure of molecules? Kekulé's formulas for organic compounds provided answers to these questions, and van 't Hoff's theory of the tetrahedral carbon atom extended the image of molecules into three dimensions. Thus, in the course of the long nineteenth century, chemists came to use molecular formulas not just heuristically, but as reliable indicators of the linkage of atoms in molecules. Positivism gave way to an acceptance of the validity of formulas as models and paper tools (Klein 2003).

This brought order to organic chemistry and had a bearing on inorganic chemistry. Mendeleev's periodic law and periodic table, resting on accurately determined atomic weights, brought a similar order to inorganic chemistry. Frankland's work on organo-metallic compounds could be seen as leading to a truly unified and comprehensive chemistry. The increasing internationalism of chemistry and the growth of university departments of chemistry, in both of which, by the late nineteenth century, Germany had assumed the leading role, provided a professional foundation for major industrial and economic recognition of chemical science.

Another important problem for chemists had been the nature of chemical affinity. Sometimes presented as the attraction of like substances, sometimes as the attraction of opposites, it could offer an organizing principle for reactions, but had no real explanation until the nineteenth century, when Berzelius, Davy, and Faraday offered electrical explanations. By the end of the century, chemical thermodynamics provided a more comprehensive theory of affinity. It is striking that the greatest development of chemical thermodynamics was in the work of Gibbs, a theoretician, and van 't Hoff, one of the founders of the new physical chemistry. Both emphasized studying chemical processes rather than the specifics of molecular structure, but the former could fairly be called *general* chemistry, a label especially used by Ostwald to emphasize that physical chemistry would unite all the subdisciplines of chemistry. By 1914, all these branches of chemistry – organic, inorganic, physical, and analytical – would chart their own course and be joined by new ones, such as chemical physics and theoretical chemistry. But they all remained interconnected parts of a unified chemistry, brought together in a long process that began in the eighteenth century.

CHAPTER TWO

Practice and Experiment: *Analysis, Synthesis, and Paper Tools*

YOSHIYUKI KIKUCHI

This chapter is concerned with the broad question of what chemists actually did in the nineteenth century. The use of "practice" and "experiment" in the chapter title suggests that chemists have primarily been doing experiments. That is in the main correct, but the real question is: what exactly did the act of experimentation entail in nineteenth-century chemistry? Until the late eighteenth century, chemistry had a strong association with natural history – the collection, description, identification, and classification of species in the animal, vegetable, and mineral kingdoms. Alongside this natural-historical tradition of chemistry stood the laboratory tradition of manipulating material substances. The latter could be roughly classified into two categories: analysis, or the detection, separation, and purification of materials into their simpler components, and synthesis, the construction of materials from simpler components. Since the early eighteenth century, analysis complemented by synthesis had been considered the most reliable method of determining the composition of a material as the success of the latter corroborates the accuracy of the former (Klein and Lefèvre 2007: 115). Examining influential textbooks of chemistry published in late-eighteenth-century France, the epicenter of the "Chemical Revolution," reveals a delicate balance between the natural-historical and laboratory (primarily analytical) traditions in chemistry (Bensaude-Vincent

1990). During the nineteenth century, this balance in chemical practice shifted decidedly away from natural history and toward analysis and synthesis in the laboratory which was the most important change in the culture and practice of chemistry in this period.

That said, one of the most important transformations in nineteenth-century chemical practice had seemingly little to do with the manipulation of materials in the laboratory. Following the invention and use of chemical symbols such as C, O, and H by Berzelius in 1813, chemists in the nineteenth century increasingly employed Berzelian symbols and numbers to represent the composition of chemical compounds, created visual images with these symbols, and built three-dimensional physical models as shorthand and heuristic tools to record and aid their analytical and synthetic endeavors. The manipulation of formulas and models thus formed another integral component of chemical practice, and the application of Berzelian symbols for modeling and understanding of organic compounds and reactions triggered the nineteenth-century transformation of organic chemistry from primarily a natural-historical science, with the task of identifying and describing substances found in animals and plants, to an experimental science that created and analyzed carbon-containing compounds in the 1840s (Klein 2003: 4–5).

Other important non-material tools used by chemists in the nineteenth century include the periodic table employed by Mendeleev, and numerical tables, graphs, and mathematical equations used by late-nineteenth-century physical chemists such as Arrhenius and van 't Hoff. Chemical practice in the nineteenth century therefore consisted of both the material culture of measuring instruments, chemical apparatus, the chemical substances themselves, and the increasing number of various kinds of "paper tools" for understanding the chemical transformations found in the laboratory and for designing further experimental investigations on the bench.

ANALYSIS

The eighteenth century already saw considerable developments in both qualitative and quantitative chemical analysis. One such example was blowpipe analysis, which employed a curved tube through which the experimenter blew a stream of air. By directing the air through the flame of a candle or a burner, chemists could create a tiny jet of intense heat that would otherwise be available only in large furnaces. The blowpipe had existed since antiquity; its chemical use was pioneered in the seventeenth and eighteenth centuries by German scientists (Bud and Warner 1998: 68). In eighteenth-century Sweden, however, the blowpipe was redesigned as a smaller, simpler, and portable tool. As a result, it came to be commonly used by chemists and mineralogists there and contributed to the isolation of a number of elements such as nickel and manganese (Figure 2.1; Dolan 2003; Abney Salomon 2019). The isolation of elements by blowpipe analysis was a precondition for the articulation of the analytical or operational concept of the chemical element by Antoine-Laurent Lavoisier, according to which an element is defined as "the last point which analysis is capable of reaching" (Lavoisier 1789; McEvoy 1988).

Likewise, Lavoisier's "balance sheet" method was based on his own and other chemists' quantitative analyses. In 1792, Jeremiah Richter coined the term "stoichiometry" to describe the relative proportions of chemical compounds in their reactions (Burns et al. 2014). Combined with the chemical atomism of Dalton and Berzelius, volumetric and gravimetric stoichiometric calculations became the experimental basis of elemental analysis and the kernel of chemical practice in the nineteenth century.

Although there was continuity between the practice of analysis from the eighteenth century to the nineteenth century, the apparatus and equipment used by chemists underwent a significant change. For example, to convince his audience that the results of his analysis were sufficiently precise, Lavoisier would often have large, complex, and expensive apparatus constructed by instrument makers. For example, the 1785 gasometer built for the public

FIGURE 2.1 The blowpipe in use (Mawe 1825: 4). Courtesy of the Wellcome Collection.

demonstration of the decomposition of water has been dubbed the "Rolls Royce of chemical instruments" (Levere 2000: 122f; 2001: 72). The Parisian tradition of constructing large, expensive apparatus continued in the nineteenth century by Joseph-Louis Gay-Lussac and Louis-Jacques Thenard, but the overall direction in the development of chemical analysis later in the century would be to develop cheaper, more affordable options that did not depend on the expertise of instrument makers. Two of the most prominent shifts were toward miniaturization of apparatus and the greatly increased use of glassware, the "glassware revolution" (Homburg 1999; Jackson 2015b). Both trends were closely tied to the separation of the academic chemical laboratory from the artisanal laboratory or factory that originally had a shared material culture and the emergence of the teaching student laboratory in the first half of the nineteenth century (Klein 2008).

Jacob Berzelius, famed as the most accomplished analyst of the day, stood at the center of such development. Trained in Uppsala in the use of the blowpipe by Johan Gottlieb Gahn, Berzelius inherited the Swedish tradition of mineral blowpipe analysis and became its doyen in the early nineteenth century (Dolan 2003; Abney Salomon 2019). The major advantages of the blowpipe over the furnace in mineral analysis was its portability, economy, and efficiency, and Berzelius' continued preoccupation with these factors was the driving force behind his roles in both the miniaturization and the glassware revolution of chemical analysis in the early nineteenth century.

Known for his glass-blowing skill, Berzelius is credited as the inventor of the glass test tube and separatory funnel, and popularized the usage of table-top laboratory equipment such as beakers by prominently illustrating them in his textbooks and articles (Figure 2.2; Szabadváry 1966: 144–50; Jackson 2015b: 53–8). As discussed in the previous chapter, chemists would not agree on a unified set of atomic weights until the 1860s, but they invariably used the combining weight of the element with another element of known atomic weight in a binary compound as an experimental basis for determining atomic weights. Therefore, compiling an atomic weight table involved a variety of analytical methods: the detection of mineral elements within a sample by blowpipe and precipitation; electrolysis by current electricity, pioneered by Humphry Davy and Berzelius in the 1800s; the separation of a sample into different mineral substances by precipitation, crystallization, and filtration; and weighing with a sensitive balance. Armed with the blowpipe and other laboratory tools and an increasing knowledge about the properties of mineral substances (such as solubility with acids and the color of solutions and precipitates), Berzelius raised the standard of mineral analysis to the point that he produced the most accurate values of atomic weights for most elements in the 1810s.

The next generation of chemists would compile and systematize the various analytical methods in the form of textbooks. Christoph Heinrich

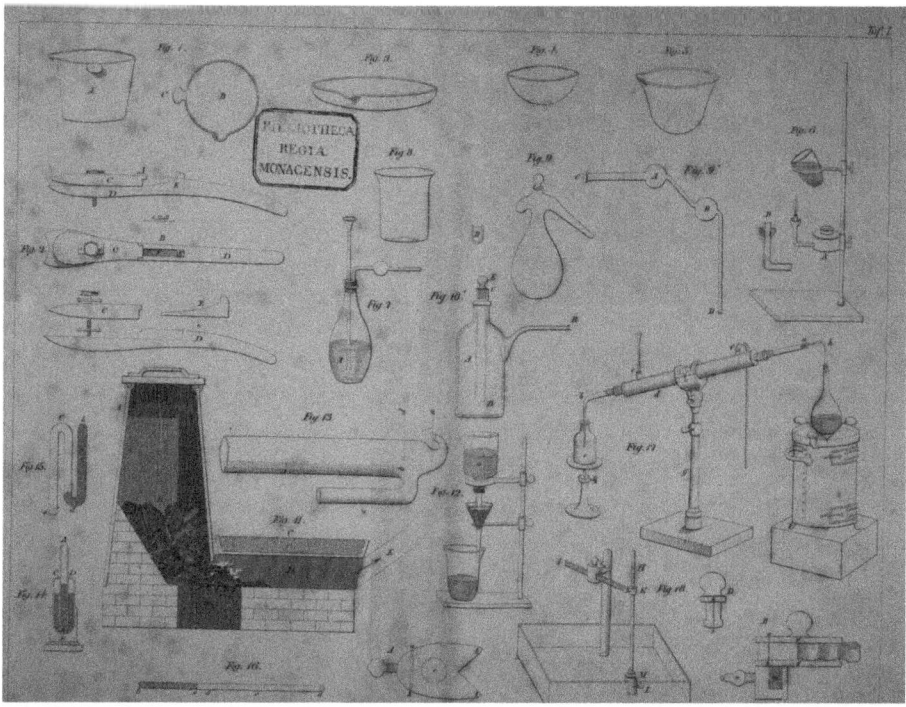

FIGURE 2.2 Berzelius' laboratory equipment (Berzelius 1841). Courtesy of the Bayerische Staatsbibliothek München, Chem. 30 d-10,a, um:nbn:de:bvb:12-bsb10072019-2.

Pfaff published the first major textbook of analytical chemistry in 1821, in which he extensively used hydrogen sulfide as a reagent (Szabadváry 1966: 150–5). Following Pfaff's lead, Berzelius' German student Heinrich Rose systematized qualitative analysis by using hydrogen sulfide and other reagents such as hydrochloric acid and ammonium sulfate as "grouping" reagents in his *Handbook of Analytical Chemistry* (1829). With "grouping" reagents, an experimenter would be able to classify and separate components in a solution into groups by their characteristic reactions with these reagents that produced colored solutions or precipitates. Tests with such reagents would become a standard part of university chemical training in qualitative analysis. One of the most widely used textbooks of qualitative analysis in the nineteenth century was Carl Remigius Fresenius' *Instructions in Qualitative Chemical Analysis* (1841), which had reached sixteen editions by the time of the author's death in 1897 and been translated into various languages (Szabadváry 1966: 168). It was a comprehensive manual with the clear presentation of the subject matter of analytical chemistry pioneered by Rose. Fresenius worked as an assistant in Liebig's analytical laboratory at Giessen, where he taught chemical analysis according to his textbook, between 1841 and 1845, when he became professor at the Agricultural Institute in Wiesbaden. In 1848 he founded his own private laboratory there, albeit partly with state sponsorship. Other courses in analytical chemistry were instituted in London, Manchester, Paris, and multiple additional locations in Germany.

There are several social factors in this mushrooming of analytical courses dating back to the late eighteenth century when, apart from the need to train mining professionals, pharmacists became interested in public health issues such as water and the adulteration of food, and medical doctors in forensic medicine (Homburg 1999: 9–18). In addition, German states took assertive measures toward drug regulation and pharmacist qualifications in the post-Napoleonic period, resulting in state examinations of pharmacists and physicians in some German states to demonstrate their practical skills in chemical analysis. A similar legal requirement for medical doctors to demonstrate evidence of chemical laboratory instruction appeared in England in the 1830s (Roberts 1976a: 17).

Interestingly, academic analytical chemistry was disconnected from industrial analysis until the 1850s due to the nature of analytical techniques (Homburg 1999: 20–5). Titrimetry (determining the concentrations of substances in a solution by gradually adding measured amounts of another solution of known concentration) had been used in chemical industry, especially in alkali manufacture, for quality control since the early nineteenth century, but academic chemists largely used gravimetry (determining weights of reactants or products) and were at first reluctant to include titrimetry in their repertoire. One reason for this neglect was a prejudice against this technique by academic chemists for its perceived industrial connotations, but another reason lay in the fact that, as the names "Berthollimetre," "Chlorimetre," "Alcalimetre," and "Sulfhydrometre" suggest, these tools were product-specific and lacked a general method. Generalizing the graduated scales on these instruments became possible by the introduction of the concept of "normal" solutions based on atomic or equivalent weights after 1840, and the accuracy and reliability were established both by the improvement of tools such as the introduction of the stopcock burette (to facilitate addition of tiny increments of reagents to a solution) and by adequate training. Titrimetry appears to have been given its full place in academic chemistry by the mid-1850s when it appeared in textbooks, for example Friedrich Mohr's *Textbook of Chemico-Analytical Titrimetry* (1855–1856). As Fresenius noted in the third edition of his *Instructions in Quantitative Chemical Analysis* (1853) (Homburg 1999: 24–5):

> Although volumetric analysis methods were indeed used in earlier times, nonetheless they were rather isolated, and were more often used for the technical measurement of contents rather than in scientific experiments; now, however, even with regard to the latter, the times are moving towards

using volumetry to obtain one's ends much more quickly than is possible with weight analysis methods, with the same degree of accuracy.

By the 1850s, the accuracy and reliability of academic analytical chemistry and the speed and simplicity of industrial analysis had converged.

The analysis of organic (i.e. animal and plant) substances posed a different kind of problem. Mineral substances consisted of combinations of two or three elements in usually simple proportions, so chemically characterizing a mineral substance meant, first of all, identifying the component elements by qualitative analysis, and, to a lesser degree, ascertaining the atomic combining ratios with quantitative analysis. The large majority of organic substances, on the other hand, are made from only four elements – hydrogen, oxygen, carbon, and nitrogen – but combine in a bewildering variety of proportions. That means the emphasis in organic analysis was on quantitative analysis, or accurately measuring the combining ratios of these elements. Most chemists in the first half of the nineteenth century agreed on the purpose and object of quantitative organic analysis, but there were two very different ways of doing it.

The earlier method was volumetric combustion analysis, measuring the volumes of gas products during the combustion of a sample. Its origin lay in the pneumatic chemistry of the eighteenth century, and was refined further by Lavoisier and Gay-Lussac (Figure 2.3), as exemplified by the latter's discovery of

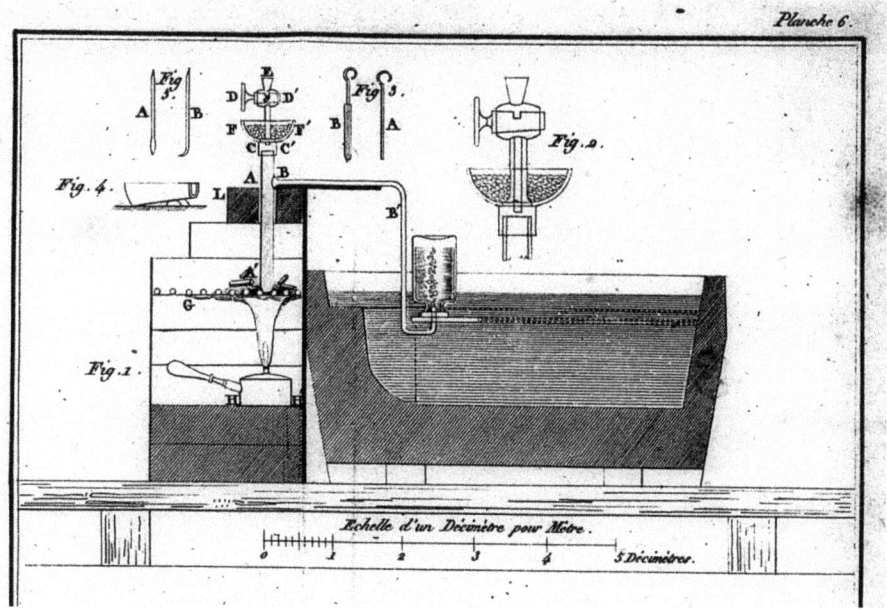

FIGURE 2.3 Gay-Lussac and Thenard's volumetric apparatus for organic analysis (Gay-Lussac and Thenard 1811). Courtesy of the Osaka Prefectural Central Library, Japan.

the law of combining gases (1808). In 1814, Berzelius dramatically transformed this large, complex apparatus into a less-expensive table-top version with new glassware he had blown himself (Figure 2.4). His motive for this innovation came partly from the fact that he was less wealthy than his Parisian colleagues such as Gay-Lussac and could not afford the expensive gas collection apparatus. For this purpose, Berzelius adopted the second method of organic analysis, gravimetric analysis, by measuring the masses instead of volumes of combustion products by absorbing carbon dioxide with a potassium hydroxide solution. It was after a meeting with Berzelius in September 1830 that Justus Liebig started to improve gravimetric analysis; this led to his invention of the celebrated *Kaliapparat* later that autumn, the method then published in 1831 (Jackson 2015b: 60f). Because a small sample of compound produced a large volume of gas, especially in the high-molecular-weight alkaloids in which Liebig was interested, there was good reason for him to follow Berzelius' lead in adopting gravimetric analysis, because a potassium hydroxide solution could absorb a much larger amount of carbon dioxide than that a volumetric analysis could measure.

The most crucial points for successful gravimetric analysis was to ensure smooth oxidation of the sample in the combustion tube, to capture all generated carbon dioxide gas, and to make sure that the whole combustion train was airtight. Liebig's five-bulb apparatus, the heart of the *Kaliapparat*, was designed and blown from glass initially by Liebig himself. The bulbs were filled with saturated potassium hydroxide solution that dissolves the generated

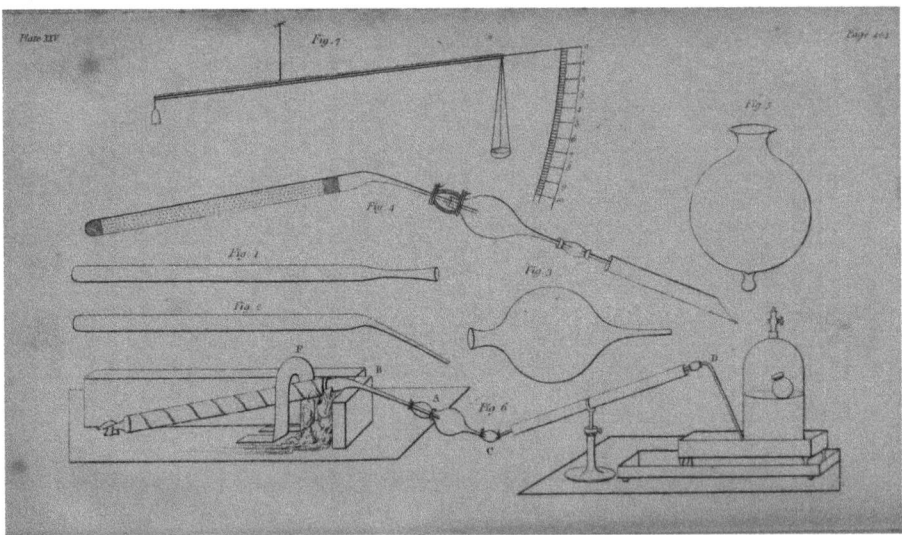

FIGURE 2.4 Berzelius' gravimetric apparatus for organic analysis (Berzelius 1814: 404). Courtesy of the Natural History Museum, London.

carbon dioxide, and the transparent glass enabled a close monitoring of bubbles, indicating the dissolution of carbon dioxide (Usselman et al 2005). After combustion, the five-bulb apparatus could simply be weighed directly to obtain the carbon dioxide content. Deceptively simple, Liebig's *Kaliapparat* was a culmination of thoughtful improvements in gravimetric analysis by means of chemists' own glass-blowing skills between the 1810s and the 1830s and became the standard method for organic analysis during the nineteenth century.

Elemental combustion analysis as described above is just part of the development of organic analysis. Organic substances must be purified, and their identities established by physical constants like melting and boiling points, refraction indices, and weights of starting materials and products, before putting their samples to elemental analysis (Ramberg 2003). Standardizing these methods of purification and identification of organic substances also has a history that stretched over the whole nineteenth century and was connected to the development of organic synthesis (Jackson 2015a). I shall discuss this aspect of analysis in the section of this chapter on synthesis.

Probably the most important innovation in chemical analysis after Liebig's *Kaliapparat* was the introduction of spectroscopy by Robert Bunsen and Gustav Kirchhoff, which enabled the quick detection of elements in extraordinarily tiny amounts (Laidler 1993, Chapter 6). Spectroscopy had distant origins in qualitative flame tests, i.e. detection of metallic salts by the colorization of the hot flame, which had been known at least since the mid-eighteenth century and was used in blowpipe analysis. William Hyde Wollaston observed the dark lines that appear in the solar spectrum in 1802, and Joseph von Fraunhofer independently rediscovered dark lines in 1814 (Usselman 2015: 72). Other pioneers in spectral analysis include William Henry Fox Talbot in the 1830s, who suggested using flame colors for qualitative analysis, and William Allen Miller in the 1850s, who used spark spectra to study spectra of many elements. The combination of spectral analysis and photography was pioneered by John Herschel and John William Draper in the 1840s and by William Crookes in the 1850s. There were, however, several problems with detecting metals using the colors of flames. It only worked for a small group of metals, only relatively cool flames were available, and small traces of sodium created a bright yellow color that would overwhelm the sample's color (Laidler 1993: 175–8).

Bunsen was in a good position to overcome these problems and to develop spectral analysis (Morris 2015: 130–5). When he established a new chemical laboratory at the University of Heidelberg in 1855, gas (as well as water) lines were introduced (the initial heat source of Liebig's *Kaliapparat* was charcoal). As discussed in the next chapter on laboratory equipment, the invention of the Bunsen burner was unthinkable without this urban infrastructure, and so was spectral analysis. No less important and fruitful was Bunsen's collaboration with his colleague the physicist Kirchhoff, who first suggested using spectroscopy to differentiate between flames with similar colors. They took great care in preparing highly purified salts as standards, used the Bunsen burner with different fuels, and constructed the first spectroscope to map the line spectra of several known elements. They then embarked on the analysis of spectra from minerals and mineral water from spas and mines, which led to the discovery of two new elements: cesium (1860) and rubidium (1861).

Once established as a reliable analytical technique, spectroscopy came to play an important role in chemistry, physics and astronomy. Many new elements were discovered using spectral analysis, such as thallium in 1861 by William Crookes and Claude Auguste Lamy, indium in 1863 by Ferdinand Reich and Hieronymus Theodor Richter, helium in 1868 by Norman Lockyer, gallium in 1875 by Emile Lecoq de Boisbaudran, and the rare gases such as argon, xenon, and neon in the 1890s by William Ramsay, Lord Rayleigh, and Morris Travers. Moreover, spectral analysis provided physicists as well as chemists with a vast amount of information on spectral lines that they tried to classify and explain. The examples best known today were the "principal," "diffuse," and "sharp" lines (the later quantum chemical terms of s-, p-, and d-orbitals come from them) classified by Cambridge chemists George Downing Liveing and James Dewar, and the empirical studies of spectral series by Johann Jakob Balmer in 1885, Johannes Robert Rydberg in 1890, and Friedrich Paschen in 1908. These spectral lines would eventually be explained by the quantum theory in the 1910s by Niels Bjerrum and physicist Niels Bohr. These cases exemplify the increasing overlap between chemistry and physics at the turn of the century.

PAPER TOOLS BRIDGING WORK AT THE LABORATORY BENCH AND THE DESK

Berzelius introduced alphabetical letters as chemical symbols in 1813 as a way of simplifying the representation of chemical compounds. A major function of Berzelian formulas was undoubtedly to represent the composition and stoichiometric relationships of chemical substances, but they soon became much more powerful tools or signs representing individual particles in combination and suggesting further investigative pathways in analysis and synthesis. (Klein 2001; Klein 2003).

While the Berzelian formulas were similar to today's chemical formulas, Berzelius used a plus sign in a formula in a slightly different way from today's notation, to show that a chemical compound consists of two or more groups of elements. For example, he wrote the formula for copper sulfate, a salt, as $SO_3 + CuO$ to show that copper sulfate has a binary constitution composed of what contemporary chemists considered sulfuric acid (SO_3) and copper oxide (CuO) and that these two elemental groups are stable enough, at least in principle, to be isolated by chemical analyses. In short, Berzelian formulas were binary constitutional formulas based on Berzelius' electrochemical dualism (cf. Chapter 1 in this volume) and at the same time heuristic "paper tools," made by rearranging the symbols in the formula, that would enable inferences about the internal arrangement of atoms within a compound. These inferences would then direct further (successful as well as unsuccessful) analytical endeavors on the bench to isolate such groups. These functions of Berzelian formulas as "paper tools" would be made apparent in the context of the development of the practice of organic chemistry.

As discussed in the previous section, organic quantitative analysis underwent vast improvement in the first three decades of the nineteenth century. The importance of composition and empirical formulas (derived by compositional data and the measurement of molecular weights) for contemporary chemists is underscored by the dispute between Liebig and Friedrich Wöhler in 1824 concerning their analyses of silver fulminate and silver cyanate, two compounds that have quite different properties but apparently had identical chemical formulas ($AgC_2N_2O_2$), suggesting that one of their compositions must be incorrect if the composition alone were responsible for their properties (Brock 1997: 72). They could solve the dispute only by doing the analysis together and concluded that both compounds did indeed have the same elemental composition. In 1830, this and other examples led Berzelius to suggest that these compounds should be called "isomers," meaning two or more organic compounds that have the same empirical formula but different properties (Rocke 1984: 173). This made acute the need to elucidate the internal arrangement of atoms in organic molecules.

The celebrated 1832 joint paper by Wöhler and Liebig, "Researches Respecting the Radical of Benzoic Acid," was one of the early examples to use Berzelian formulas as "paper tools" to infer the binary constitution of an organic compound (Brock 1997: 78f; Jackson 2008). Their strategy was to submit bitter almond oil (today's benzaldehyde) to various reactions and do an elemental analysis of the oil and each product. After analyzing these compounds, they concluded

> that they all group around one single compound, which does not change its nature and composition in all its combining relations with other bodies. This stability, this consequence of the phenomena, induced us to consider that body is a compound base and therefore to propose for it a peculiar name, i.e. benzoyl. The composition of this radical we have expressed by the formula 14C + 10H + 2O. (Benfey 1963: 34)

It is important to note here that Wöhler and Liebig did not isolate this "compound" or "radical" in the laboratory. It merely existed on paper as the common pattern (14C + 10H + 2O) that was made to appear by Wöhler and Liebig in every formula in the above network of reactions. It was the power of Berzelian formulas as paper tools, together with Wöhler and Liebig's innovative usage of them that made their argument convincing. Berzelius indeed welcomed their conclusion and in 1833 postulated a theory of binary constitution for organic compounds, the so-called "radical" theory (Klein 2003: 156–61).

Because "radicals" within molecules are primarily found by manipulating Berzelian symbols, different chemists often had very different ideas about which atomic groupings constituted real radicals, sometimes leading to controversies. Disputes over the proper arrangement of atoms became even more intense when the confidence in the radical theory started to erode in the late 1830s with the emergence of the substitution theory of Liebig's French rival, Jean-Baptiste Dumas. According to this theory, electronegative chlorine, for example, could take the place of electropositive hydrogen, which explicitly contradicted Berzelius' electrochemical dualism. Again, Dumas' concept of "substitution" exemplified the power of Berzelian symbols as paper tools, because Dumas did not isolate all the reaction products, but interpreted the reaction process on paper (Klein 2003: 191–9). It is important to note that these seemingly theoretical issues of molecular constitutions had repercussions for experimental practice. According to contemporary chemists like Liebig, the determination of empirical formulas was accomplished not merely by following protocols for using the *Kaliapparat* but also by exercising subtle judgment as to which data to select, which is inseparable from the analyst's empirical experiences, *and* theories of constitutions (Jackson 2008: 41).

Chemists increasingly relied on paper tools throughout the nineteenth century. At first, these were simple Berzelian formulas, followed by "type" formulas, August Kekulé's "sausage" formulas (Figure 1.2), Alexander Crum Brown's bond-line formulas (Figure 1.3), and by the end of the century, Jacobus Henricus van 't Hoff's tetrahedral carbon atom (Figure 1.5), to name just a few. One obvious example of the greater role of paper tools for chemists is the symmetrical benzene ring (the cyclohexatriene formula) postulated in 1865 by the principal founder of organic structural theory, German chemist August Kekulé (Figure 1.4; Rocke 2010). When drawn on paper, Kekulé's ring formula enabled him to predict that there should be only one isomer of a mono- and penta-substituted derivative of benzene, and three isomers (*ortho-*, *meta-*, and *para-*) of a di-substituted derivative. The experimental confirmation of his predictions in the late 1860s and 1870s vastly increased the attraction of organic structural theory and the power of paper tools.

A parallel trend in mid-century involved manipulating three-dimensional models of molecules in doing chemistry and thinking about the microworld. The two most famous of these models are the two-dimensional ball-and-stick models of August Wilhelm Hofmann (Figure 2.5) and the three-dimensional

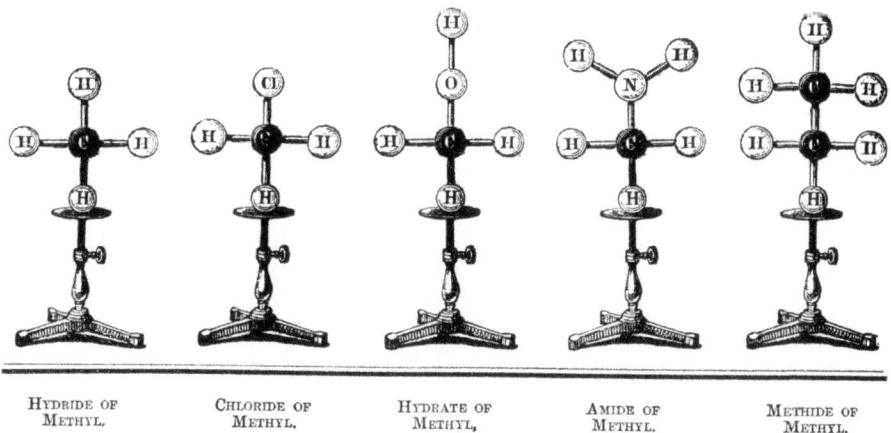

| HYDRIDE OF METHYL. | CHLORIDE OF METHYL. | HYDRATE OF METHYL, | AMIDE OF METHYL. | METHIDE OF METHYL. |

FIGURE 2.5 Hofmann's "glyptic" (sculpted) models (Hofmann 1865: 426). Courtesy of the Oesper Collection, University of Cincinatti.

ball-and-wire tetrahedral models developed for lecture demonstrations by Kekulé (Figure 2.6). These models were originally created for pedagogical purposes and not as the physical representation of molecules, and chemists warned strongly against interpreting them physically, i.e. as something showing the actual positions of atoms in a molecule. However, once this practice of using 3D models gained a foothold in the 1860s, some chemists began to use them as a heuristic tool for research (Klein 2001; Ramberg 2003; Ramberg 2014).

The crucial step of interpreting the tetrahedron as a graphic literal representation of the arrangement of valances around the carbon atom was taken by van 't Hoff in 1874 (Figure 1.5), ushering in the new field of stereochemistry (Ramberg and Somsen 2001; Ramberg 2014). Chemists such as Johannes Wislicenus, Victor Meyer, Arthur Hantzsch, and Emil Fischer utilized such spatial arrangements and van 't Hoff's theory as an explanatory device to explain isomerism, i.e. to differentiate isomeric compounds that were indistinguishable under Kekulé's organic structural theory. Conversely, they also used spatial arrangements as an isomer-counting device to predict the number of isomers, and the experimental confirmation of these predictions naturally increased the appeal of stereochemistry. In this sense, the role of van 't Hoff's theory in chemical practice resembled that of Kekulé's benzene ring, that is to establish the structural identity of isomers, predict new stereoisomers, and explain existing isomeric relationships (Ramberg 2003: 322f). This stereochemical practice in organic chemistry was also incorporated by Alfred Werner in his new coordination theory of inorganic compounds during the 1890s.

Implicit within the use of these various modes of representation is the so-called hypothetico-deductive method: proposing a hypothesis, deducing the

FIGURE 2.6 Kekulé's tetrahedral models of six carbon atoms forming a benzene ring. Courtesy of the University of Ghent.

logical consequences from this hypothesis and submitting them to experimental tests in order to confirm or refute the hypothesis (Rocke 1985; Rocke 1990; Rocke 2018b). A good early example of the self-conscious usage of this method is Alexander William Williamson's elucidation of the formula for ether. Ether is formed in a relatively simple procedure by heating alcohol in the presence of sulfuric acid. The reaction forms both ether and water. The reaction is form of dehydration, so the rational formula for alcohol could be $C_4H_{10}O \bullet H_2O$, where alchohol is a "hydrate" of ether, and the water is simply removed to leave ether with the formula $C_4H_{10}O$. Williamson, however, explained the reaction by using an alternate set of atomic weights, according to which alcohol had the formula C_2H_6O.

Williamson devised a chemical reaction that would help to decide between the "hydrate" formula and his own. If the formula was the "hydrate," the reaction would yield two products. His own formula, however, predicted only a single reaction product called a "mixed ether." When he did this reaction, he obtained a single product, vindicating his own formula. Williamson subsequently applied the same kind of reasoning to another class of organic compounds (the ketones), and during the 1850s, other chemists adopted Williamson's method to ascertain the formulas for other classes of organic compounds (Rocke 2010: 22–7). Williamson also approached the problem of etherification in a novel, graphical way, where it took place in a dynamical manner.

Williamson's methodology was well recognized by one of his students, the Japanese chemist Joji Sakurai, who commented on his mentor's research on etherification (Kikuchi 2013: 67):

> [As] a theoretical chemist, he was one of the greatest men of his time. He and Gerhardt did more than any others to bring about a complete revolution in Chemistry, by doing away with the Equivalent System which was full of confusions and inconsistencies, and firmly establishing, in its stead, the Molecular Atomic System which has ever since been universally adopted. It was, more particularly, Dr. W.'s discovery of the so-called "mixed ethers" that gave a final blow to the Equivalent System. There was nothing particularly wonderful about the experimental part of this ever memorable work of Dr. W's, but it was the keenness of his insight that was so remarkable and the logic of his argument that was so convincing.

As clarified by Sakurai, at the center of his mentor's argument lay an experimental outcome refuting one theoretical position and establishing another. This is the hallmark of the hypothetico-deductive method.

Kekulé, van 't Hoff, and Williamson all used different kinds of formulas – structural, tetrahedral, and type formulas – but were clearly bound together by common methodological threads. They are characterized by the effective use of paper tools, visualization techniques, and the hypothetico-deductive method. Thus, a significant part of the process of ascertaining the structure of organic compounds, the major goal of nineteenth-century chemists, was done not on the laboratory bench, but at the desk.

Thus far, we have focused on paper tools in the practice of organic chemistry, but these paper tools were also applied in inorganic chemistry, in the creation of the periodic table in 1869 by Dmitrii Mendeleev (Gordin 2018). What made his 1869 periodic table an effective paper tool is the balance between its theoretical commitment and flexible consideration of empirical data. While using atomic weights as a basic criterion for sorting elements, Mendeleev also used chemical properties such as oxidation levels. If the positioning of one element based on its atomic weight was at odds with that based on chemical combination, he trusted the latter and did not hesitate to revise atomic weights and change the order of chemical elements, especially when this value was not deemed certain. He also considered the list of chemical elements available then as incomplete, which enabled him to imagine the existence of new elements and predict their chemical and physical properties. As in the case of organic structural theory and stereochemistry, the isolation of Mendeleev's predicted elements – especially gallium (Mendeleev's "eka-aluminum") in 1875, and germanium (his "eka-silicon") in 1886 – played a large role in the acceptance of his table. Mendeleev's periodic table as a theoretical construct was thus malleable, neither axiomatic nor finished, and worked at the threshold of theoretical and experimental practices and of explanatory and predictive enterprises. That is exactly why it was so successful; and it tells us something essential about not only the periodic table itself, but also about the centrality of paper tools and three-dimensional models in the chemists' toolbox.

SYNTHESIS

The spread and development of chemical formulas and hand-held models outlined above also had the effect of stimulating chemists' desire to prepare organic compounds, both synthetic and natural. It is amply demonstrated by the episode of William Henry Perkin, the inventor of mauve, the first synthetic coal-tar dye. In 1856 Perkin had tried to prepare quinine by oxidizing allyltoluidine, using chemical formulas as a guide (Travis 1993: 36). He failed to make quinine, but serendipitously found a commercially successful dye. Perkin's attempt at synthesizing a target molecule was but one of the many failed experiments by chemists in the latter half of the nineteenth century; successful syntheses were difficult and rare even after the advent of organic structural theory in the late 1860s. Although structural formulas did not necessarily facilitate the synthesis of organic compounds, synthetic organic chemistry is still widely considered to have been born in this period. How might one resolve this apparent paradox?

One way of resolving this issue is to recognize two different kinds of "synthesis" in the nineteenth century. One is "constructive synthesis," i.e. the construction of a target molecule from simpler components that we usually understand by the term "synthesis" (today this is called "total synthesis"). The other method is "synthetical experiments," intended to be exploratory and analytical rather than constructive, which was introduced in 1845 by August Wilhelm Hofmann and Sheridan Muspratt in their study of toluidine (Jackson 2014a; Jackson 2014b). Hofmann and Muspratt's synthetical experiments evolved from the methodology of their teacher, Liebig, in his joint paper with Wöhler on bitter almond oil discussed above. Whereas Liebig and Wöhler's strategy was to perform a series of reactions such as oxidation and substitution on a starting reagent to identify its radical, Hofmann and Muspratt's goal was to isolate and describe the properties of the products formed when starting reagents were subjected to a much wider range of standard reactions (Jackson 2014a: 67). These standard reactions included oxidation/reduction, nitration (inserting a nitro group into the starting material), halogenation (inserting a halogen atom such as chlorine and iodine), alkylation (inserting a hydrocarbon group), and, finally, the elimination of atoms from the molecule as exemplified by the reaction that became known as the "Hofmann elimination." The crucial point about synthetical experiments was that Hofmann and Muspratt used this method primarily for analytical purposes, i.e. to get information about the constitution and reactivity of the starting materials. It was not about the construction of target molecules per se.

To be sure, synthetical experiments were prerequisite to constructive syntheses, as they provided information about how starting materials reacted under certain conditions, which in turn became part of chemists' toolkit for successful constructive syntheses. A notable example of such experiments was Carl Graebe and Carl Liebermann's 1868 recognition that alizarin, the sought-after natural red dye from madder root, is structurally related to anthracene, a coal tar constituent (Travis 1993: 171–2). This outcome suggested to them that anthracene would be a promising starting reagent and pointed to a viable pathway for the synthesis of alizarin. Their work almost immediately resulted in the commercially successful preparation of artificial alizarin, which eventually destroyed madder cultivation.

In most cases, however, the amount of chemical information derived from synthetical experiments was only a necessary but not a sufficient condition of a successful constructive synthesis. That is because the outcome of synthetical experiments became unpredictable as the structures of starting materials became more complicated (in alkaloids, for example), for these reactions produced complex mixtures of new unknown compounds, often with extremely similar properties. These compounds would then need to be experimentally separated, isolated, and identified in a reliable way for those reactions to be utilized later in constructive synthesis (Ramberg 2003; Jackson 2014a).

Devising a reliable way to identify chemical compounds was also essential to determine whether the synthetic and naturally occurring molecules were identical. The measurement of boiling points, melting points, and other physical constants such as refraction indices and optical activity for the purpose of their identification were introduced over the course of the nineteenth century (Jackson 2015a). In the 1820s, French chemist Michel Eugène Chevreul's investigation of fats included pioneering measurements of their melting points as an indicator of their chemical identity and purity, but he continued to use other properties such as smell, taste, appearance, consistencies, and origin, following the natural-historical tradition of chemistry, keeping these properties on an equal footing with physical constants. His attempt at establishing the melting point as an indicator of chemical identity did not attain general currency at that time, partly due to the lack of interest by other chemists, and because there was no standardized method of measuring melting points.

Hermann Kopp's extensive research between the 1840s and 1860s, first in Giessen and later in Heidelberg, on numerical relationships between the chemical compositions and boiling points of organic substances in the same homologous series was mutually stimulated by Hofmann and Muspratt's synthetical experiments, and raised chemists' interests in measuring physical constants. As Kopp collected boiling point data for different substances from different chemists, he became frustrated with the fact that there was no operational criterion for boiling points, and therefore suggested various ways of standardizing measuring procedures, including now-familiar practices such as placing a thermometer above the boiling liquid at a consistent height (rather than submerging its bulb in the liquid) and putting platinum wire within the liquid to avoid "bumping" of the liquid directly into the condenser.

Still, establishing boiling points and melting points as criteria for identification remained elusive for Kopp, and required further efforts and skills of both academic and industrial chemists such as Ludwig Gattermann (who published an influential manual of laboratory practice for chemists in 1894), and glassblowers like Heinrich Geissler (who provided purpose-made apparatus for melting point measurement and specialist thermometer) until the turn of the century (Jackson 2015a: 200). The need for chemists such as Albert Ladenburg (who in 1886 first synthesized coniine, an alkaloid poison extracted from hemlock) to identify a large number of new compounds created by their syntheses was the major impetus for establishing boiling points and melting points as criteria for identification, because the traditional criteria of identification such as smell, taste, appearance, crystal shape, and physiological properties proved insufficient to differentiate new materials. Glassblowers played a particularly important role in the standardization of boiling and melting point measurements, as they manufactured precision thermometers for chemists who in turn recommended these thermometers in their writing, thus promoting a de facto standardization of thermometry (Jackson 2015a: 203).

Constructive synthesis also relied heavily on new methods for the separation of different compounds from complicated mixtures. The two pillars of separation techniques of chemists in the nineteenth century were fractional crystallization and fractional distillation. Fractional crystallization, which required the painstaking control of experimental conditions, observation of different crystal forms, and sometimes good fortune, was pioneered by Chevreul in his research on fats and oils, and effectively used by Ladenburg and other chemists for their synthetic endeavors. Distillation techniques have a very long history dating back to antiquity, but they had serious problems when applied to organic chemistry, partly because the heat required for distillation often destroys delicate organic substances, and partly because many natural organic liquids, such as petroleum, are mixtures of compounds with similar boiling points. These problems had been recognized by the late eighteenth century, but were not solved until the latter half of the nineteenth century, with the advent of fractional distillation techniques, along with reduced-pressure distillations and vacuum distillations (Holmes 1971; Holmes 1989a; Jackson 2015a). The improvement and commercial availability of vacuum pumps partly explains this timing, but the impact of urban infrastructure in chemical laboratories, most importantly the indoor water supplies that became available in many Western cities by mid-century, was essential. A laboratory equipped with running water could make use of simple water aspirators to create modest low pressures; these were used to accelerate distillations and filtration of crystals (Morris 2015).

THE EMERGENCE OF PHYSICAL CHEMISTRY AND ITS IMPACT ON CHEMICAL PRACTICE

As the career of Hermann Kopp suggests, there were much deeper connections between organic chemistry and physical chemistry at the level of practice throughout the nineteenth century than has been hitherto recognized. In fact, Kopp was once a professor of physical and theoretical chemistry at Giessen, and his thermometric research outlined above is considered an early prominent example of the "physicalist tradition in nineteenth century chemistry," a precursor of physical chemistry (Servos 1990: 11–20). Other good examples include Victor Regnault, renowned for his precise thermometric studies of gases as well as his synthesis of chlorinated hydrocarbons, and Heinrich Buff, who published the *Textbook of Physical and Theoretical Chemistry* (1857) together with Kopp and Friedrich Zamminer (Rocke 2018b: 580–1). However, until the late nineteenth century the emphasis of chemical practice was on analyzing reaction products, ascertaining their structures, and synthesizing target molecules, with the measurement of a compound's physical properties often used only as a tool for these purposes.

Exceptionally early cases of mathematically treating chemical reactions and their physical conditions as a main topic from the mid-century could be found in the studies of reaction rates or chemical kinetics (King 1981; King 1984; Laidler 1985; Laidler 1995). Apart from its role in the subsequent development of physical chemistry, the measurement of time spent in chemical processes also had practical industrial applications because, as one of the founders of physical chemistry, Wilhelm Ostwald, succinctly put it, "it is important to speed up slow reactions as much as possible since for industrial chemistry, as for any other, time is money" (King 1981: 70). The research on the rate of inversion of sugar using a polarimeter by Ludwig Ferdinand Wilhelmy, published in 1850, is usually considered as the first contribution in this field, but was virtually unnoticed by contemporaries. Wilhelmy was also the first to use differential equations in following the course of chemical reactions, the practice developed by later kineticists.

In the 1860s, the works of two groups, Cato Maximillian Guldberg and Peter Waage in Christiania (now Oslo), Norway, and Augustus George Vernon Harcourt and William Esson in Oxford, almost simultaneously established the laws of reaction rates. The methodology of Guldberg and Waage was informed by an analogy with classical mechanics following the tradition of chemical affinity (King 1981: 71). According to their reasoning, in the state of equilibrium (in which a chemical system is at rest) two opposing forces, one causing a forward reaction and the other a reverse reaction, are balanced. Guldberg and Waage then crucially considered these driving forces as proportional to the velocity of reactions, which led to the first mathematical statement of the law of mass action ("the rate of a chemical reaction is proportional to the product of the concentrations of reactants") in 1864 (King 1981: 73). Analogies with mechanics were also important for Harcourt and Esson, but their approach is best characterized by its experimental rigor in choosing suitable model reactions and in utilizing the techniques of titrimetry, and their full recognition of time and temperature factors as crucial to the understanding of chemical reactions, which eventually led to the elucidation of the temperature dependence of reaction rates by van 't Hoff and Arrhenius (King 1981: 74–80; King 1984).

Another area in which the quantitative treatment of reactions and their conditions became paramount from the mid-century is the study of heats of reaction, or thermochemistry, pioneered by Julius Thomsen from the 1850s and Marcellin Berthelot from the 1860s (Kragh 1984). Again, the better understanding of chemical affinity by means of quantification was the main motive of their research, because the principle of energy conservation, established by the early 1850s, suggested using the amount of heat evolved by the formation of a chemical compound as the measure of its chemical force or affinity. Thomsen and Berthelot's painstaking measurements with instruments such as a "mixing calorimeter" or bomb calorimeter covered most groups of organic and inorganic substances. Their theoretical commitment to Thomsen's principle ("every simple or complex action of a purely chemical nature is accompanied by evolution of heat") or Berthelot's Principle of Maximum Work ("every chemical change accomplished without the intervention of external energy tends to the production of that body, or system of bodies, which disengages most heat") later drew criticism from proponents of physical chemistry but Thomsen and Berthelot's very attempt to measure chemical affinities by physical quantities gave inspiration to other chemists to redirect the course of chemistry (Dolby 1984; Kragh 1984: 257, 260).

As the end of the nineteenth century approached, these mathematical and physical methods to elucidate chemical reactions gradually became part of the toolkit of chemists (Servos 1990; Laidler 1993). An early sign of this sea change is Wilhelm Ostwald's claim at the examination for his master's degree at the University of Dorpat (now Tartu, Estonia) in 1877: "Modern chemistry is in need of reform." Ostwald called for chemists to redirect their attention from the substances entering chemical reactions to the reactions themselves, especially their rates and equilibria (Servos 1990). He was inspired by the thermochemical studies of Thomsen and the research by Guldberg and Waage that led to the law of mass action in the 1860s. Because precise calorimetric measurements needed sophisticated equipment and it was difficult to directly measure the concentrations of participating substances in a reaction at equilibrium (as this very act of measuring disturbs the state of equilibrium), Ostwald incorporated the measurement of a variety of physical properties such as volumes, refraction indices, and electric conductivities into his project.

Ostwald's interest in electric conductivity brought him into contact with Swedish chemist Svante Arrhenius, whose 1884 Ph.D. dissertation was on the electric conductivities of a variety of electrolytes. Ostwald's research on chemical equilibria and reaction rates had a similar effect on his friendship with van 't Hoff, who had switched his research interest from stereochemistry to chemical equilibrium and wrote *Études de dynamique chimique* (1884). This book was van 't Hoff's magnum opus in physical chemistry, a thorough mathematical treatment of reaction rates and chemical equilibria based on thermodynamics. The collaboration of these three chemists brought about a new discipline of physical chemistry based on the concept of ions, charged particles that were supposed to exist in solutions that caused their electrical conductivity.

Toward the end of the century, the impact of the new physical chemistry on chemical practice was considerable. Along with some familiarity with theoretical and experimental physics and the technique of precision measurement, the very nature of physical chemistry encouraged the frequent use and manipulation of numerical tables, graphs, and mathematical equations that required calculus. In arguing for the formation of ions in his 1887 paper on dissociation, Arrhenius never claimed to have isolated ions, but drew up numerical tables containing the degrees of dissociation, i, calculated from the measurement of osmotic pressure and those calculated from electric conductivities. While admitting the existence of some exceptions, Arrhenius claimed that "[there] appears nevertheless on a comparison of the numbers in [two columns of the table] a very marked parallelism between them. This shows a posteriori that in all probability the assumptions on which I have based the calculation [i.e. van 't Hoff's law of osmotic pressures and the ionic theory] are in the main correct" (Nye 1984: 243f). For Arrhenius, showing good numerical agreements was the essential part of chemical argumentation.

That chemists in the nineteenth century were on the whole unfamiliar with higher mathematics is shown by the occasional collaborations of a chemist and a mathematician in the early history of physical chemistry, including the above examples of Guldberg and Waage and Harcourt and Esson in the 1860s. By the late nineteenth century, physical chemists were required to master certain levels of mathematics and physics, and this expectation was slowly extended to chemists in general throughout the course of the twentieth century.

It is misleading, however, to assume that the rise of physical chemistry created a total transformation of chemical practice during the nineteenth century. Rather, it gave rise to a cultural divide between physical chemists and chemists trained in the traditional methods of synthesis and analysis, especially organic chemists. Henry Edward Armstrong was an outspoken critic of Arrhenius and "ionist" physical chemistry; he disliked the ionic theory partly for cultural reasons. In Armstrong's own words (Dolby 1976: 389):

> The fact is, there has been a split of chemistry into two schools since the intrusion of the Arrhenic faith, rather it should be said, the addition of a new class of worker into our profession – people without knowledge of the laboratory arts and with sufficient mathematics at their command to be led astray by curvilinear agreements; without the ability to criticise, still less of giving any chemical interpretation. The fact is, the physical chemists never use their eyes and are most lamentably lacking in chemical culture. It is essential to cast out from our midst, root and branch, this physical element and return to our laboratories.

The comparison of Armstrong's and Arrhenius' quotations makes it clear that physical chemists were of a different kind, trained in a different set of practices, unlike those in organic or analytical chemistry. It would take another half-century or so for physical and organic chemistry to come together with the emergence of physical organic chemistry in the mid-twentieth century, signified by the publication of the influential textbook of this new discipline by American physical chemist Louis Plack Hammett in 1940.

CONCLUSION

In this chapter we have asked the questions: what did chemists do in the nineteenth century, and how did what they do change over time? Overall, the center of gravity in chemical practice shifted from a natural-historical science of collection, description, and classification to an analytic-synthetic science employing both laboratory apparatus and theoretical constructs as "tools." The former tools included the *Kaliapparat* and other types of glassware, burners, spectrometers, precision thermometers, and galvanometers. The latter tools consisted of chemical notation, two- and three-dimensional models, tables of numbers and symbols, and mathematical equations. These "paper tools" worked well due to their constant dialogue with experimental data and to a wider use of the hypothetico-deductive method among chemists. Experiments and theories in nineteenth-century chemistry became so mutually intertwined that it is sometimes almost meaningless to categorize these practices as either "experimental" or "theoretical."

Some elements of the transformation of chemical practice were external to the science. For example, the development of urban infrastructure such as gas and water lines helped establish spectral analysis, fractional distillation, and crystallization as viable laboratory techniques. Many others, however, came from within chemical practice. The difficulty chemists encountered in analytically separating individual chemical species from mixtures created by their synthetic endeavors necessitated the measurement of physical constants such as boiling and melting points as criteria for identifying chemical compounds, which in turn provided an impetus for the development of the new field of physical chemistry. These mutual stimuli between elements of chemical practice had the effect of multiplying and diversifying chemists' toolkit in the nineteenth century.

This transformation, however, did not create a single coherent epistemic or cultural system, but rather a methodological and cultural pluralism in chemical practice, which could be roughly divided into organic, inorganic, analytical, and physical chemistry. Overemphasizing these divisions is misleading because strong (although sometimes strained) connections always existed between these subdisciplines, and, conversely, there were different styles of doing research in a single subdiscipline (for example, electrochemical dualism versus the type theory in organic chemistry). This pluralism may seem impure, parochial, or opportunistic, but it was remarkably fruitful in elucidating the structures of organic and inorganic compounds and quantifying and controlling chemical reactions (Rocke 2018b). The issue of reconciling these subfields with each other, and indeed chemistry with other scientific disciplines such as physics and biology, provided chemists in the twentieth century with further opportunities and challenges.

CHAPTER THREE

Laboratories and Technology: *Continuity and Ingenuity in the Workplace*

AMY A. FISHER

INTRODUCTION

In the opening pages of his popular textbook *The Elements of Experimental Chemistry* (1815), the physician and chemist William Henry wrote that "in chemical changes, we may always observe an important difference in the properties of things; their appearance and qualities are completely altered, and their individuality destroyed." He defined chemistry as "that science, the object of which is to discover and explain the changes of composition that occur among the integrant and constituent parts of different bodies" (Henry 1815: xix–xx). He asserted that while having a formal "chemical laboratory, though extremely useful, and even essential, to all who embark extensively in the practice of chemistry," it was "by no means necessary." All a chemist really required was access to a furnace and a well-lit room, "easily ventilated, and destitute of any valuable furniture" (Henry 1815: 1).

In contrast, by the end of the nineteenth century, Paul Freer, professor of chemistry at the University of Michigan, emphasized that "chemistry [was] growing to be more of an exact science every day" and "what [was] *essential* in the science ... [was] *quantitative*" laboratory research. "Certain kinds of *good* apparatus every chemical laboratory *must* have in order to do successful work," he argued, included "a good balance, a barometer, two gasometers for the storing of larger quantities of gases, and a combustion furnace" (Freer 1895: iii, original emphasis). His detailed "Laboratory Appendix" filled more than a quarter of the whole book, educating interested teachers and students about safety precautions, such as when to use a fume hood (or fume cupboard), and providing detailed instructions on how to use a thermometer, eudiometer, battery, Bunsen burner, and sophisticated glassware for chemical analysis. With proper attention to detail, Freer noted, "a chemical experiment, when correctly performed, is as certain to have an unvarying result as is one in physics" (iv).

These differences between Henry's and Freer's description of the necessary equipment for doing chemistry are representative of the significant changes that occurred in chemical standards, instrumentation, and laboratory design and construction during the nineteenth century. Late eighteenth-century chemical practitioners often worked in laboratories in private residences, in institutes of higher education as affiliates to university medical programs, or in artisanal workshops, such as apothecary shops and assay offices (Klein 2008). Although few could afford expensive precision instruments, many practitioners routinely repurposed household items, made their own glassware, and/or built specialized laboratory equipment for their experiments (Golinski 1992; Rocke 2000b; Jackson 2015b). They focused on understanding the relationship between chemical composition, physical properties (e.g. weight and volume), and chemical behaviors (e.g. many acids reacted with metals to form salts). Their research tended to emphasize the study of airs, acids, minerals, and inorganic materials with only a few elementary components (Klein and Lefèvre 2007).

By the late nineteenth century, however, many chemical laboratories had migrated from small, privately owned spaces to institutes of higher education and large-scale industry (Homburg 1999). Instruments that had been handcrafted earlier in the century became standardized and mass-manufactured, and organic chemistry flourished with the widespread adoption of German chemist Justus Liebig's system of teaching and research (Dyer and Gross 2001; Morris 2002; Rocke 2003). Chemists also took advantage of developments in urban infrastructure during this period – i.e. the construction of city water, sewer, and gas mains and, later in the nineteenth century, electric power – to revamp the structure of their laboratories, improving safety and expanding their research capacity. Laboratories became "matter factories," sites that produced not only knowledge of chemical processes but also novel chemical products (Morris 2015).

Despite innovations in laboratory design, there was also significant continuity in chemical technology and practice during this period, as chemists continued to use furnaces, thermometers, distillation apparatus, and color indicators in their research. Yet, the development of new instruments – such as the *Kaliapparat* (or five-bulb apparatus), spectroscope, and polarimeter – allowed chemists to tackle the composition and structure of more complex organic substances than ever before. As the number of precision instruments in analytical organic chemistry increased, chemists developed deeper insights into stoichiometry and stereochemistry (Rocke 1984; Rocke 2010).

A small group of chemists, however, were dissatisfied with the emphasis that organic chemists placed on the composition and structure of substances. Wilhelm Ostwald, for example, argued that "modern chemistry [was] in need of reform" (Servos 1990: 3). He asserted that by studying what happened as reactants *became* products, chemists could develop more general laws of chemical change. In this growing subfield – physical chemistry – chemists modified existing instruments, such as the calorimeter and electroscope, and developed new laboratory techniques to determine reaction rates and measure other important thermal and electrochemical properties (Ostwald 1912: iv; Servos 1990). Although the study of hitherto unknown substances and energy in the late 1890s further complicated chemical theory and practice, chemistry and its three major branches – inorganic, organic, and physical – were well established in institutes of higher education by the early twentieth century.

This chapter explores three accompanying developments in chemical laboratories and technology over the course of the nineteenth century. First, it describes the establishment, growth, and sophistication of laboratories used for teaching and research at universities. Second, it emphasizes technological developments in inorganic and analytical organic chemistry: what motivated chemists to develop new technology and how in turn did these new techniques and instruments affect chemistry? Third, it focuses on instruments that used heat to investigate and analyze chemical change, including furnaces, blowpipes, and newer technology such as the spectroscope and the *Kaliapparat*. Yet by the end of the century, older, seemingly mundane equipment continued to be not only relevant, but essential amid methodological and technological innovation. Chemistry also became increasingly quantitative, as the standards for what constituted good experimental research became more rigorous.

THE EVOLUTION OF THE LABORATORY: FROM PRIVATE WORKSPACES TO LARGE-SCALE TEACHING–RESEARCH LABORATORIES

Eighteenth-century laboratories were "places for making things" (Figure 3.1). Sketches of the leading French chemist Antoine-Laurent Lavoisier's laboratory emphasized large tables and pneumatic apparatus for the study of gases as

FIGURE 3.1 An engraving by Alfred William Warren (1822) depicting a laboratory designed mostly for solitary work. Courtesy of the Science History Institute, Philadelphia.

well as walls "lined with pigeonholes for the storage of apparatus" and neatly arranged jars of chemicals (Morris 2015: 39). His privately owned laboratory and instrument collection, cataloged after his death, listed approximately 7,000 instruments built by more than seventy different artisans. It included, for example, a precision balance, a gasometer, and a pneumatic machine made by Nicolas Fortin, a leading French instrument maker (Beretta 2014). With these hand-crafted instruments, Lavoisier became adept at measurement and brought attention to hitherto imperceptible effects. He published quantitative studies, especially of changes in weight, volume, pressure, and temperature during chemical reactions, and "stressed the accuracy of the procedures and the high standard of proof [he had] thereby achieved" in chemistry (Golinski 1994: 94).

Changes in higher education further facilitated the dissemination of Lavoisier's work. Established in 1794, the year of Lavoisier's execution for crimes against the state that he ardently denied, the *École Centrale des Travaux Publics*, renamed the following year the *École Polytechnique*, aimed to inculcate its students with an understanding of science and technology in a three-year degree program. Although the institution focused on mathematics and civil and military engineering, it employed three distinguished chemists: Jean-Antoine Chaptal, who promoted Lavoisier's theory of acidity and combustion, and Antoine-François de Fourcroy and Claude-Louis Berthollet, who had collaborated with Lavoisier to construct a new system of chemical nomenclature. In addition to learning chemistry, enrolled students also had the opportunity to participate in small group discussions and to perform experiments themselves with the guidance of a professor and/or his laboratory assistant. Some experimental equipment came from Lavoisier's own laboratory, instruments seized at the time of his arrest, as well as from the Military Engineering School at Mézières. Students also learned glass-blowing and made apparatus in-house. This marked an important shift in science education. In the past, professors would lecture and perform demonstrations in auditoriums while enterprising students could seek further training in chemistry from instructors with private laboratories, if they so desired. At the *École Polytechnique*, however, everyone was expected to "get their hands dirty" and learn about and participate in experimental research (Bradley 1976: 426–33).

Although the school's novel approach to science education encouraged student enrollment and foreign visitors, it failed on many levels to achieve its goals. With the outbreak of the Napoleonic Wars in 1803, military recruitment increased and funding for laboratory courses decreased. Plans to expand the existing laboratory facilities also failed, creating a crisis for chemistry instructors. Fourcroy complained about the lack of running water, and he and others worried about the poorly ventilated laboratories (Bradley 1976: 438–41).

Many institutions faced similar challenges in incorporating laboratory chemistry into the curriculum, but a number of different schools embraced the *École Polytechnique* model of small recitation sections and teaching–research laboratories. Founded in 1802, the United States Military Academy (USMA) transformed its curriculum in 1817. After receiving degrees from Dartmouth College and USMA and two years of study at the *École Polytechnique*, its new superintendent, Major Sylvanus Thayer, developed a rigorous curriculum that emphasized the importance of not only military training but also the study of science and engineering, especially algebra, trigonometry, physics, and chemistry. Other schools in America in the 1820s also emphasized the importance of laboratory work, often drawing on the *École Polytechnique* for inspiration, such as the Rensselaer School (later renamed the Rensselaer Polytechnic Institute) and the University of Virginia (Angulo 2012).

In the 1820s, a teaching–research laboratory was built in the University of Virginia's iconic Rotunda, a building designed by Thomas Jefferson, the third US President and founder of the university (Kelly 2015). Initially, John Emmet, the first professor of natural philosophy and later chemistry at the college from 1825 to 1842, had been allocated a small room in the basement for chemical experiments, but he lobbied Jefferson for more space and equipment for students, adopting his former advisor William MacNeven's hands-on approach to teaching chemistry (Emmet 1826; John G. Waite Associates 2017: 38–56). MacNeven, chair of chemistry at the College of Physicians and Surgeons in New York City, insisted "students ... perform chemical experiments instead of merely witnessing them" (Stookey 1965: 1040; Miles 1970). Aware of pedagogical trends in France and as "an ardent utilitarian in chemistry," Jefferson accepted many of Emmet's suggestions (Greene 1984: 170; John G. Waite Associates 2017). In a semicircular recess in the basement builders constructed the laboratory's brick-and-mortar furnace along with five student workstations, fireboxes for burning wood and coal, numerous flues, underground ductwork for ventilation, and a rudimentary fume hood for venting noxious gases as well as cupboards for storing chemicals (John G. Waite Associates 2017: 75–81). As with Thayer's curricular reforms, Emmet's laboratory design illustrated his commitment to chemistry education (John G. Waite Associates 2017: 37).

In 1826, the year after the University of Virginia started offering courses in chemistry, changes in science education were also under way in Europe. After studying at the universities of Bonn and Erlangen in the early 1820s, Justus Liebig completed his training in Paris. At the *École Polytechnique*, he attended the chemistry lectures of Joseph Gay-Lussac, who succeeded Fourcroy as professor of chemistry in 1810, and Louis Jacques Thenard, a *répétiteur* at the *École Polytechnique*, professor of chemistry at the *Collège de France*, and lifelong friend of Gay-Lussac (Holmes 1989b: 124–5). Returning home in 1824, Liebig accepted a position as professor of chemistry at the University of Giessen (Brock 1997: 37–71).

Before he traveled to Paris, Liebig dreamed of establishing a teaching–research laboratory similar to Johann Trommsdorff's Chemico-Physico-Pharmaceutical Institute in Erfurt (Holmes 1989b: 124). His experience in Paris reinforced this goal, but much to his frustration, in Giessen he received little financial support and inherited meager facilities. His laboratory, housed in a decommissioned military barracks, consisted of a small lecture room, "an unheated broom closet that served as a balance room and store," and a room filled with furnaces and tables that could accommodate approximately eight students on the ground floor (Brock 1997: 47). Liebig lived with his family in a small apartment on the second floor. Like Emmet, Liebig had to purchase most of the chemicals and sundry supplies and pay his assistant's salary out of his own modest stipend (Brock 1997: 59). His early unsuccessful attempts to persuade administrators to support his laboratory reflected the early-nineteenth-century conflict in German-speaking universities over not only the goals of higher education but also who those institutions served. University officials argued that the purpose of higher education was to prepare civil servants for public life, not to train technicians and artisans for employment (Holmes 1989b: 127).

Despite these setbacks, Liebig's institute underwent small expansions between 1826 and 1832 (Morris 2002: 92–6). In 1832, he petitioned the university to renovate his laboratory space. He noted the disagreeable and unsafe conditions in which he and his students worked, e.g. the main workspace lacked proper ventilation, which exposed everyone in the laboratory to hazardous fumes. He requested that the existing classroom be reconfigured into a private laboratory and weighing room, for precision balance measurements, and that a new lecture hall be constructed out of the storage room. Although the university approved his proposal, construction was not completed until 1835 (Holmes 1989b: 146–7).

Yet, students flocked to work with Liebig. His approach to chemistry was exciting and systematic, enabling students to master fundamental procedures before engaging in independent research. He also developed novel experimental techniques and instruments, such as the *Kaliapparat* (discussed below), to analyze complex organic substances and explore isomerism (Rocke 2003). As Liebig's prestige grew, additional modifications to his laboratory further enhanced its reputation. Working with architect Paul Hofmann, Liebig received government funds to expand his facilities, providing space for separate pharmaceutical and chemical laboratories as well as a large auditorium, and to incorporate innovations such as glass-fronted fume cupboards, which allowed experimenters to monitor chemical reactions (Brock 1997: 57–8). By 1847, Liebig proudly wrote that his laboratory was "no longer the damp, cold, fireproof vault of the metallurgist, nor the manufactory of the druggist, fitted up with stills and retorts." Rather, students worked in "a light, warm, comfortable room" (Figure 3.2; Liebig 1847: 8). Gas burners had largely replaced coal-burning furnaces, eliminating the stench of smoke and reducing respiratory illness. With these improvements, he wrote, "health is not invaded, nor the free exercise of thought impeded; [in this laboratory] we pursue our inquiries, and interrogate nature to reveal her secrets" (8).

The development of Liebig's teaching–research laboratory not only exemplified the "experientialist pedagogical philosophy" emerging during this period, but also demonstrated an important shift in state funding and university policies supporting chemistry in German-speaking regions (Rocke 2003: 106). Influential inorganic chemist Robert Bunsen also experienced this shift. As a student, he worked at the University of Göttingen in Friedrich Stromeyer's laboratory, which included portable furnaces, a separate room for precision instruments, and laboratory space for student workstations. At the Polytechnic School in Kassel where Bunsen taught from 1836 to 1839, however, cramped conditions forced students to perform experiments wherever they could find space, e.g. in the furnace room and schoolyard. Poor ventilation in the actual laboratory had corroded physical instruments and negatively affected faculty and student health (Nawa 2014: 127). Although Bunsen moved on to better and more prestigious positions at the University of Marburg and the University

FIGURE 3.2 Lithograth of Justus von Liebig's analytical laboratory in Giessen, *ca.* 1845. Contrast with the laboratory of Figures 3.1 and 7.1. Courtesy of the Sidney M. Edelstein Library, The National Library of Israel.

of Breslau, his laboratory working conditions remained poor (Jackson 2011: 56). Thus, when Bunsen was offered the chair of chemistry at the University of Heidelberg in 1852, he brokered the construction of a new building for his laboratory as a condition for accepting the offer. He insisted the new building had to have good ventilation and rooms for not only laboratory experiments but also a lecture hall, a workshop for making instruments and blowing glass, a library, chambers for gas analysis and electrochemical research, separate storage spaces for chemicals and instruments, as well as a small private laboratory, office, an apartment for Bunsen and his family, and living quarters for his laboratory assistants. Baden's Ministry of the Interior consented to Bunsen's demands, in part because of the promise that chemical studies offered for improving local agriculture and industry, broadly construed (Nawa 2014).

Bunsen's new facility was impressive. Completed in 1856, the lecture hall contained 110 seats arranged in ten ascending tiers along with equipment for in-class demonstrations, including a large table and a "bench-mounted fume cupboard" as well as access to running water, electricity, and gas lines. The laboratory for advanced students had twenty-eight individual workstations. It was connected to a slightly smaller laboratory with twenty-two workstations for beginners. Both rooms were equipped with oak workbenches. Each station had storage space for students and a small vent as well as access to a chemical cupboard, a gas line, fresh water, and electricity, while wastewater was collected in oak barrels. In between the two laboratories, common equipment was stored, such as furnaces and hot plates. Adjacent to the advanced-student space was Bunsen's own private and well-lit laboratory, which included four workbenches, a sink, and numerous precision instruments and glass apparatus (Nawa 2014: 130). In short, each laboratory was equipped with state-of-the-art technology.

Although chemistry was gaining more recognition and prestige as a discipline in the early- to mid-nineteenth century, many chemists at British and American universities did not have access to the same resources as their German counterparts. Between 1820 and 1840, for example, British natural philosophers and chemists urged reform, yet Oxford University resisted updating its curriculum to include the experimental sciences, in part because of its strong tradition in natural theology, its own self-described mandate to teach classics, and its student clientele – approximately 80 percent of Oxford undergraduates became clergymen during this period (Crosland 2003: 402). Although Oxford hired Charles Daubeny as professor of chemistry in 1822, he also later served as the university's professor of botany and rural economy. Initially given a small space for a laboratory in the basement of the Old Ashmolean Museum, Daubeny asked for permission to create a laboratory in the Physic Garden attached to Magdalen College; he paid for half of the costs himself. Like Liebig's colleagues at Giessen, Oxford administrators viewed chemistry as a physically dangerous manual activity that did not belong with the strictly intellectual activity of the university, and Daubeny struggled to secure support (Crosland 2003: 405–10).

After years of lobbying, in the early 1850s, the university acquiesced in the Royal Commission's recommendations to improve science education at Oxford. The institution hired Daubeny an assistant, Nevil Story Maskelyne. While Daubeny provided chemistry demonstrations in his private laboratory in Magdalen College, Story Maskelyne designed and admitted undergraduate students to a course on experimental chemistry in the Old Ashmolean Museum. The university also allocated funds from its profitable University Press to construct a new science building (Rowlinson 2009: 93–6).

After Daubeny resigned his chemistry lectureship in 1854, Henry Acland, professor of medicine, worked with the architects Benjamin Woodward and Thomas Deane on the new facilities. Acland proposed using the architectural template of the Abbey of Glastonbury to create a natural history museum and chemistry laboratory, the latter modeled after the Abbot of Glastonbury's kitchen. The historic kitchen appealed to Acland, with its high ceilings and four chimneys (one at each corner), and fit in with the university's gothic architecture. Neither Acland nor the architects, however, considered recent developments in laboratory design, and the new building, with its insufficient lighting and poor ventilation, was less than ideal for chemical research (Crosland 2003: 413). Newly hired professor of chemistry Benjamin Brodie, who had studied with Liebig at Giessen and succeeded Daubeny at Oxford, requested additional funds to purchase chemicals and connect the laboratory to the local gas lines and sewer system. Although he initially failed to convince the council of the necessity of these costs, he continued to request improvements to the space. In the 1860s and 1870s, as the number of chemistry students continued to grow, the university consented to some of Brodie's proposals. In 1877, five years after Brodie's retirement, the "kitchen" was finally renovated and construction started on a new building adjacent to the Museum, which included a large laboratory for inorganic and analytical chemistry as well as smaller rooms for organic chemistry (Rowlinson 2009: 93–103).

While Daubeny and Brodie worked to persuade university officials of the importance of chemistry during this period, many Oxford Colleges developed their own independent programs to support scientific inquiry. Donations by physicians John Freind and Matthew Lee in the eighteenth century helped to establish the Christ Church laboratory for the study of anatomy and chemistry, and Balliol College set up a basement laboratory in the Salvin Building for chemical experiments for its students. In 1874, Balliol further partnered with Trinity College to hire additional lecturers and to construct a new facility for the natural sciences, including chemistry, which reflected the increasing demand for and importance of the experimental sciences in higher education (Rowlinson 2009: 113–23).

In Massachusetts, Eben Horsford initially received more support from his colleagues than Daubeny and Brodie. A graduate of Rensselaer Polytechnic Institute, Horsford also studied with Liebig at Giessen before being appointed the Rumford Chair of chemistry at Harvard University in 1847. Although Harvard president Edward Everett secured a $50,000 donation for a new chemistry laboratory, those funds were divided, much to Horsford's dismay, with $25,000 allotted for Horsford's laboratory and $25,000 earmarked for a new engineering facility as well as faculty positions in geology and engineering (Rossiter 1975: 75). With insufficient funding for staffing and maintenance, Horsford's attempt to model his laboratory after Liebig's failed, and it became a more modest venture.

In contrast, Horsford's colleague Josiah Parsons Cooke Jr., a mathematics tutor and self-taught chemist, was appointed Erving Professor of chemistry and mineralogy in the Harvard University Medical School in 1850 after his predecessor, John Webster, was convicted of homicide. Located in the basement of University Hall, near the college bakery, Cooke's first research laboratory at the institution did not have access to running water or natural gas (Jensen 2011: 12). Like Brodie at Oxford, Cooke managed to create a vibrant curriculum and research program for determining atomic weights largely because he could not safely perform organic chemistry experiments in the allocated space. Eventually, in 1857, Cooke successfully lobbied the administration not only to construct a new chemistry building and laboratory, separate from Horsford's facility, but also to transfer his faculty position from the Medical School to Harvard College. With this institutional reorganization, undergraduate courses in chemistry became mandatory and the nascent program grew from a single faculty member – Cooke – to a fully fledged chemistry department by 1894, "servicing over 315 students and boasting of three full professors, three instructors, eight assistants, more than 16 course offerings, and a graduate program" (Jensen 2011: 11).

As these examples show, in many different settings, chemistry flourished despite significant infrastructural challenges. In the last third of the nineteenth century, however, there were significant changes to university teaching–research laboratories as most of the facilities constructed prior to 1866 became obsolete in German-speaking regions as a result of the rapid industrialization of chemistry. Scholars have estimated that in 1860 "there were some 3,000 'known' carbon compounds, by 1899 there were 74,000, with millions more still open to investigation" (Johnson 1985: 104). With the race to isolate and identify potentially lucrative new chemicals, teaching and research laboratories became larger and more complex and afforded more safety features and amenities.

Nowhere was this change more apparent than in the laboratories directed by August Wilhelm Hofmann. Hofmann studied in Liebig's laboratory and in 1845 was appointed as the director of London's newly created Royal College of Chemistry. Open to men and women from a wide variety of socioeconomic backgrounds, the college provided Hofmann with a good salary and benefits, but provided little laboratory equipment. Students had to purchase, for example, their own test tubes, flasks, beakers, rubber tubing, and crucibles. They also routinely complained of shortages of reagents and more complex distillation apparatus nominally provided by the college. The basement of the chemistry building had storage space for some equipment, such as thermometers, autoclaves, and furnaces, as well as a mixing room and a small kitchen that doubled as a laboratory for preparing reagents. The ground floor included a library, two balance rooms, Hofmann's office and private laboratory, as well as a meeting room and small lecture theater. The top floor housed the main laboratory with forty-six benches. It was, however, so poorly ventilated that some experiments had to be conducted on the roof when the weather allowed (Gay and Griffith 2017: 18).

Despite these challenges, Hofmann built up a successful research program on aniline chemistry (Brock 2008; Jackson 2011). As with Liebig's laboratory, Hofmann's students made significant contributions to chemistry and the chemical industry, including "Charles Blachford Mansfield, who pioneered the commercial distillation of benzene (but was burnt to death as a result); William Perkin, who founded the synthetic dye industry; and William Crookes, who discovered thallium and set up *Chemical News*" (Morris 2015: 155). Buoyed by this success and the political changes occurring in Europe, Hofmann left

London for Berlin in 1865, enticed by a professorship in chemistry and the promise of a new laboratory, where he worked to strengthen ties between the government, the chemistry community, and chemical industries.

Hofmann's former pupil Crookes wrote in 1869 that Hofmann's gigantic new research building in Berlin was like "a mansion of a noble kind – a palace" (Morris 2015: 146). Filled with well-stocked laboratory benches with access to water and gas as well as fume cupboards and sinks, it was a veritable display of cutting-edge technology. The building was also connected to the newly built city-wide sewage system, introduced in the 1850s to prevent the contamination of the water supply and reduce outbreaks of cholera (Morris 2015: 149–50).

The layout of the laboratory also dramatically changed from earlier in the century. Benches were no longer placed along the central axis of the room but at the periphery, effectively creating a central aisle through the space. Benches continued to be made out of wood (e.g. pine or teak) with flame-retardant materials such as asbestos mats protecting the paraffin-wax sealed surfaces. Hofmann also used "gas burners to draw air from the main laboratory through the [fume] closet," a forerunner of the fume hood, dramatically improving air-flow and ventilation (Jackson 2011: 58). There were also a number of specialized spaces for different kinds of instrumentation. Darkrooms, separate from the main laboratory, were constructed for optical experiments. Because some reactions involved highly volatile substances, Hofmann also incorporated ballistic cabinets – often housed in rooms with reinforced walls and doors – to protect experimenters from accidental high-temperature and pressure explosions (Jackson 2011: 59–60). Other rooms were specifically dedicated to delicate precision instruments and gas analysis to reduce environmental sources of error and contamination (Morris 2015: 155–65).

Hofmann's large, well-equipped laboratory, however, was by no means the only one in the 1860s. Before deciding on Berlin, Hofmann had also been offered the chair of chemistry at the University of Bonn and designed a spacious new laboratory for that institution. When he chose Berlin over Bonn, the position and facility was offered to August Kekulé, whose groundbreaking studies of aromatic compounds provided the basis for structural chemistry (Rocke 1993: 54; Morris 2015: 148). Hermann Kolbe, who also made significant contributions to organic chemistry, had an even larger and more modern laboratory than Hofmann or Kekulé (Rocke 1993). From his inadequate laboratory at Marburg, he moved to the University of Leipzig in 1865, which at the time was undergoing a renaissance. Paul von Falkenstein, minister of culture in Saxony, allocated significant financial resources to revitalizing science on campus, creating new positions and buildings for laboratory work with the support of the monarch and legislature. Kolbe received a luxurious new laboratory and residence. His three-storey, 4,660 square-meter building contained forty-four well-lit and ventilated rooms. Heated with coal-fired steam, the facility had access to running water and gas for both lighting and burners (Rocke 1993: 279–80). Kekulé's and Hofmann's lavish new laboratories at Bonn and Berlin could accept sixty to seventy students each, whereas Kolbe could accommodate up to 130 students in his monstrous new facilities (Rocke 1993: 280). While Liebig's and Bunsen's laboratories each possessed elements of these new "chemical palaces," these teaching–research laboratories were state-of-the-art research institutions.

As organic chemistry prospered, many chemists working in inorganic and the relatively new field of physical chemistry struggled to achieve the same level of prestige and institutional support for their research. Wilhelm Ostwald, who received one of the first professorships in physical chemistry at Leipzig in 1887, argued that despite organic chemists' success, their emphasis on chemical classification had not addressed key questions regarding the rate and yield of chemical reactions. Ostwald, however, faced challenges in pursing his research. In contrast to organic chemists' facilities, a reporter for *Nature* described Ostwald's laboratory as "an old pile ... the light was bad, the rooms unventilated, the heating effected by means of stoves difficult to regulate and producing dust which caused much injury to the finer instruments." The columnist concluded, "it would have been difficult to construct a laboratory worse adapted for physico-chemical investigations" (H.N. 1901). Yet, by the end of Ostwald's second year, thirty students had enrolled in his laboratory courses, a testament to his ability as a lecturer and a researcher (Servos 1990: 46–53).

Ostwald's acrimonious critiques of organic chemistry, however, often sowed discord rather than consensus, but his emphasis on measuring chemical affinity – the degree to which one substance will combine with another – rehabilitated interest in thermochemical and electrochemical instrumentation (Coffey 2008). Alluding to Lavoisier's achievements in the late eighteenth century, Ostwald suggested that chemists "not only adopt physical tools and canons of precision" but also "focus on the quantifiable aspects of chemical phenomena" (Servos 1990: 4; Kim 2006). It was not enough, he argued, to just use existing chemical apparatus. Rather, the technology itself needed to be grappled with and understood at a more fundamental level in order to develop a deeper understanding of chemical reactions and the mathematical relationships that existed between chemical variables (Ostwald 1905: iv). As laboratories grew larger and the number of students and professional chemists increased, how did chemical instruments and apparatus change? The next section examines the types of technology used at the bench.

CONTINUITY AND INNOVATION: EXAMPLES OF CHEMICAL APPARATUS

In his popular 1827 textbook *Chemical Manipulations*, Michael Faraday wrote that the key to setting up a functional laboratory was to convert a ground or basement floor for easy access to a furnace, running water, and wherever possible, sewage drains (Faraday 1827: 12; James 2010). Furthermore, he recommended that "the place should be furnished with several flues" and have adequate ventilation and lighting. He also suggested that "when a furnace ... stands against the wall, it is frequently advantageous to construct a wooden hood over [it], to receive the fumes evolved during the digestions and solutions made upon it and to conduct them away to a [separate] chimney" (Faraday 1827: 13–14). Keeping the workspace as free as possible from contaminants was crucial, he argued, not only for obtaining accurate results but also for safety (16).

In contrast to astronomy or natural history, Faraday emphasized that nature had to be coerced through assiduous chemical experimentation to reveal its inner workings (i). Thus, chemists needed to be able to reliably and accurately separate materials to control the number, kind, and weight of reactants, and to determine the type and number of reactants in a product. Most chemical reactions also involved heat. Controlling the temperature of materials similarly required technical skills and artisanship. These skills could not be easily learned through reading, only by practice (iii).

In the 1820s, Faraday emphasized the importance of the furnace to laboratory work. He took pains to describe several different kinds of portable and fixed furnaces and blowpipes, which he thought were crucial to the chemist's workspace. For small laboratories, he recommended a table-top furnace referred to as a "blue pot," made out of lead or clay and crystalline graphite, that could be used for distillation and small-scale smelting (Marcet 1817: 304–5; Faraday 1827: 84). These pots could be "easily cut by a saw, rasp, or file" and were sold at an affordable price. Faraday recommended boring holes in the side of the pot for mounting cast-iron handles and to help control the air flow to the fuel, usually coal (1827: 84–5). He also suggested that experimenters purchase a small cast-iron grate to fit inside the furnace. Changing the height of the grate relative to the fire and plugging the holes with soft brick stoppers allowed the operator to increase or decrease the temperature of the fire (86–7).

Unlike the blue pot, built-in or fixed furnaces were more complicated to build and expensive to install. William Brande oversaw the construction of the Royal Institution's fixed furnace. It was large, approximately fifty-seven inches long and forty-two inches wide. Faraday wrote that "the principal part of this furnace [was] necessarily of brick-work, only the top plate with the [sand] baths and the front, being of iron." At the back of the furnace, the wall was "at such a distance from it as to leave space for the ash-hole and fire-place; these walls [were] lined with Welch lumps," an insulating material similar to plaster (Marcet 1817: 304; Faraday 1827: 90–1; Nicholson 1845: 57).

Controlling the temperature of portable and fixed furnaces was crucial for facilitating the melting and boiling of substances, one key set of physical characteristics used to isolate and identify specific materials (Figure 3.3; Russell 2000a). In the 1840s Hermann Kopp's experimental efforts to determine whether a numerical relationship existed between composition and boiling point encouraged chemists to develop standards for measuring the melting and boiling points of substances and to collaborate with glassmakers. Boiling points could be especially difficult to determine as they depended on composition, pressure, and the physical location of the thermometer relative to the boiling material (Figure 3.4). Because of these challenges, quality glassware became increasingly important in chemical instrumentation (Jackson 2015a: 187–92).

Instrument makers, especially glass-blowers, gained prestige during this period as chemists themselves often specified in correspondence and publications the artisans responsible for the specialized glassware used in their experiments. In the 1830s, Liebig, for example, recommended Parisian instrument maker Charles Félix Collardeau du Heaume to collaborators. A decade later, Kopp continued to endorse Collardeau thermometers, which he used in his boiling

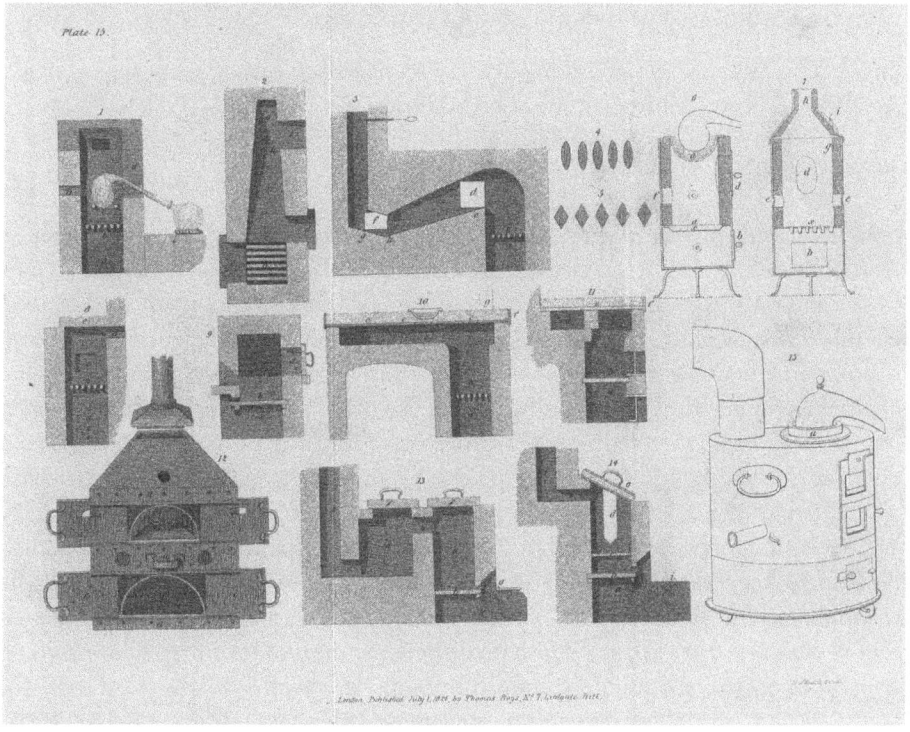

FIGURE 3.3 An illustration of assorted furnaces, including fixed and portable furnaces (Accum 1824). Courtesy of the Science History Institute, Philadelphia.

FIGURE 3.4 An etching by Walter Ouler (1875) depicting a chemist watching over a distillation process. Courtesy of the Science History Institute, Philadelphia.

point measurements, whereas British chemists preferred thermometers manufactured by Heinrich Geissler and Franz Müller (Jackson 2015a: 202).

In addition to measuring the melting and boiling point of substances, chemists also used the blowpipe (see Figure 2.1), which by the early nineteenth century had become an indispensable qualitative tool for not only identifying chemical composition but also illustrating the principles of combustion. Usually made out of brass, it consisted in its most basic form of a thin tube to direct the chemist's breath to the flame heating a substance, or to adjust the temperature of the fire under a small crucible or glass tube containing reactants (Faraday 1827: 109; Jensen 1986; Dolan 1998). Students, Faraday wrote, needed to be able to keep their mouths and cheeks full of air while learning to breathe through their nostrils – a well-known technique called circular breathing – so that they did not prematurely fatigue or become dizzy. Once they had mastered control of their breathing, the mouth became "a closed but distended bag" for trapping and propelling gases into the flame (Faraday 1827: 111–13). More sophisticated versions of the blowpipe used different-sized funnels and apertures to vary the speed and cross-sectional area of the air flow and often came with a condensing chamber to reduce the amount of moisture from the experimenter's breath injected into the flame (Miller 1856: 639–42).

Faraday suggested that students practice with a low-sitting candle or oil lamp so that they could sit with their elbows resting on a laboratory bench to help stabilize the blowpipe and control the flow and direction of their breath. He encouraged students to observe how the color and direction of the flame varied and to learn how to manipulate "the situation and powers of [the different] parts of the flame by operating on a globule of tin about the size of a piece of shot placed in a small cavity of charcoal." Faraday noted that students should observe the metal closely throughout this exercise. Depending on the temperature of the flame, "the metal will [either] be converted into a white crusty oxide or be reduced and appear in the metallic state as a brilliant fluid globule, according to the part of the flame directed upon [the metal] and the skill of the operator" (Faraday 1827: 116).

In addition to using the blowpipe as an educational tool, Benjamin Silliman Jr., professor of chemistry at Yale University, wrote that when used properly, "the Mouth Blowpipe converts the flame of a common lamp or candle into a powerful furnace" (Silliman Jr. 1856: 275). By heating small samples and observing both the physical changes to the sample and the color changes that occurred at different temperatures in the flame, the blowpipe could be used to qualitatively analyze the composition of different substances. British chemist and science educator Robert Galloway encouraged students to carefully learn the colors associated with different burning materials, e.g. "soda and its salts impart to the exterior blowpipe flame an intense yellow color" whereas "potash and its salts tinge the outer blowpipe flame violet" (Galloway 1858: 40).

The blowpipe remained important for determining chemical composition throughout the nineteenth century because of its educational value, affordability, and portability (Jensen 1986). Commercial chemistry kits in Britain and the United States often included a blowpipe, and blowpipe manuals remained popular throughout the century, as illustrated by the success of German chemists Karl Plattner and H. Theodor Richter's *Manual of Qualitative and Quantitative Analysis with the Blowpipe*, which by 1885 had gone through five editions in America (Turner 1983: 215–16; Landry 2014).

Although blowpipe analysis allowed researchers to qualitatively identify elements by the color of the flames they produced, chemists endeavored to study more rigorously the conditions under which a burning substance produced light as well as the kinds of light it radiated. In the early nineteenth century, both British natural philosopher William Hyde Wollaston and German optician Joseph Fraunhofer noted that the refraction of sunlight through a prism produced not only the characteristic solar spectrum but also many hitherto unknown fine black lines interspersed throughout it. Using the spectroscope to analyze the colors produced by flames, Bunsen and Gustav Kirchhoff showed that these lines produced in the solar spectra correlated with specific lines generated by various chemical elements, making the spectroscope a novel tool for chemical analysis (McGucken 1969).

The spectroscope (Figure 3.5) consisted of a trapezoidal-shaped wooden box "blackened on the inside" and mounted on a stand (Kirchhoff and Bunsen 1860: 90). It contained a centrally located prism on a brass plate, rotatable about its vertical axis, and two small telescopes installed midway on the inclined sides of the box. The light from the burning material traveled through the first telescope to the prism where it was refracted. The second telescope, farthest from the light source and aligned with the prism's focal point, had its optics replaced by a plate with a single slit. "By turning the prism round, every colour of the spectrum [was] made to move past the vertical [slit] of the [second] telescope, and any required position in the spectrum [was] thus brought to coincide with this vertical line." The relative location of each spectral line could then be measured by recording the color and number of the line as well as its position (Kirchhoff and Bunsen 1860: 90–1; McGucken 1969). This design was

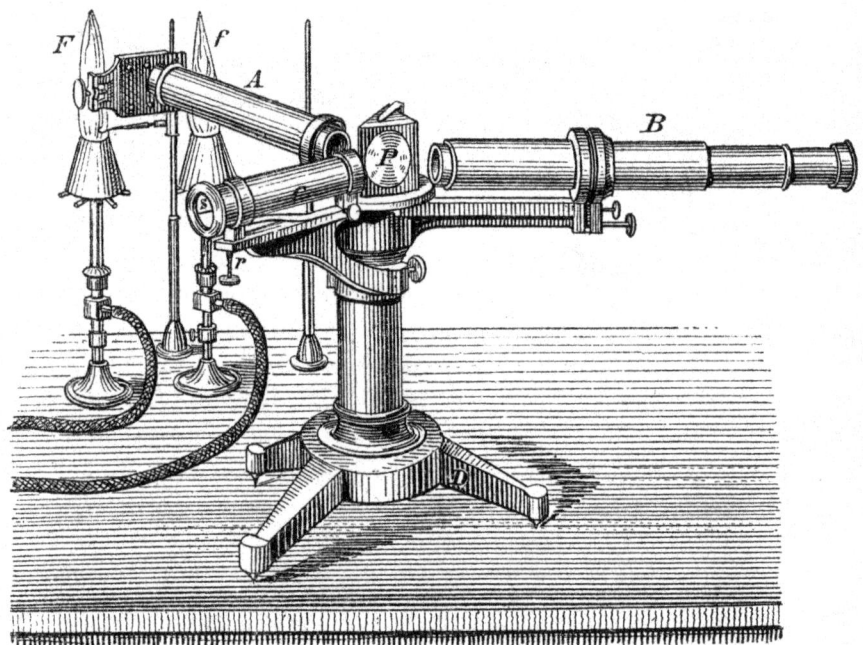

FIGURE 3.5 The spectroscope invented by Bunsen and Kirchhoff. One tube allowed light from the combustion reaction to enter the prism apparatus. A telescope provided the experimenter with a view of the spectrum. A second telescope provided the experimenter with a view of a scale with which to measure the relative location of each line in the spectrum. Photograph by Photo 12/Universal Images Group via Getty Images.

further improved upon by Munich optician Carl August Steinheil, who added a collimator to focus the light and built in a scale so that each spectral line could be identified and mapped *in situ* (Hentschel 2002: 48–52). The spectrum was further enhanced by using a novel heat source with a colorless flame.

Taking advantage of the Heidelberg laboratory's gas lines and working with Henry Roscoe, Bunsen designed this special gas burner – the Bunsen burner – to further facilitate spectroscopic research. Using the principle behind British chemist Humphry Davy's safety lamp, researchers like Faraday had previously built gas burners using a fine metal gauze to prevent the flame from igniting the source. Experimenting with the length and width of the tube and keeping the premixed gas and air mixture at positive pressure, Bunsen and Roscoe were able to eliminate the need for a wire safety gauze, creating a "hotter, more concentrated flame" ideal for photochemical studies (Kohn 1950; Jensen 2005).

Having a reliable spectral fingerprint for each element turned out to be a valuable marker of chemical composition. For example, Kirchhoff and Bunsen observed that "the luminous ignited vapour of the lithium compounds gives two sharply defined lines, the one a very weak yellow line, Li β, and the other a bright red line, Li α." These lines were unique to lithium. When a small amount of "carbonate of lithium mixed with excess of milk-sugar was burnt, the reaction was visible in a room of 60 cubic metre capacity" and "minerals containing lithium, such as triphylline, triphane, petalite, lepidolite, require[d] only to be held in the flame in order to obtain the bright line Li in the most satisfactory manner." By testing a variety of different, naturally occurring metals and silicates, they reached "the unexpected conclusion that lithium is most widely distributed in nature." They observed its characteristic lines in water from the Atlantic Ocean, kelp, "the orthoclase and quartz from the granite of the Odenwald," as well as "in the ashes of tobacco, vine leaves, of the wood of the vine, and of grapes" and "in the ashes of the crops grown in the Rhineplain near Wäghausel, Deidesheim, and Heidelberg, on a non-granitic soil" (Kirchhoff and Bunsen 1860: 96–7).

The advantage, they noted, to performing spectral analysis instead of relying on a regular color test, for which "the presence of even the smallest quantity of impurity is often sufficient to destroy the characteristic colour of a precipitate," was that "the coloured bands [were] unaffected by such alteration of physical conditions, or by the presence of other bodies." Although they could not yet explain why each element possessed a unique spectrum, the "positions which the lines occup[ied] in the spectrum [gave] rise to chemical properties as unalterable as the combining weights themselves" (Kirchhoff and Bunsen 1860: 106–7). This crucial point made it possible for Kirchhoff and Bunsen to discover the elements cesium and rubidium from their distinctive spectral lines.

In addition to this powerful new tool of chemical analysis, chemists also invented new methods to determine stoichiometry, or the combining proportions of the elements in compounds. In his 1875 Faraday Lecture to the British Chemical Society, Hofmann reflected on Liebig's significant contribution to stoichiometry, proclaiming that "Liebig's is the name and figure alone fitted to stand beside Faraday in the representation of our century to future generations," and singled out Liebig's important 1831 invention: the *Kaliapparat* or "potash bulb" (Hofmann 1876: 6). This instrument, Hofmann declared, was "in its action so perfect, in its simplicity so beautiful, for the *analysis by combustion of organic bodies*, and more especially for the determination of their carbon [content] by ponderal [weight-based], instead of, as in the old method, by volumetric measurement" (Hofmann 1876: 45–6, original emphasis). In short, the *Kaliapparat* had transformed organic chemistry.

Before Liebig, eudiometers – devices that measured the volume of gas produced during a chemical reaction – played a significant role in the development of stoichiometry. Early in the 1800s, for example, Gay-Lussac demonstrated that water contains two unit volumes of hydrogen for every one unit volume of oxygen (Crosland 1978; Crosland 2003: 92–100). Using eudiometers in organic chemistry, however, could be dangerous because of the volatility of the chemicals involved. Compounds with sizable atomic weights, especially alkaloids, were also particularly difficult to analyze, as they created large quantities of gas.

The *Kaliapparat*, in contrast, could capture large quantities of carbon dioxide by weight rather than volume (Figure 3.6). As the sample burned, carbon and hydrogen combined with oxygen to produce carbon dioxide and water vapor. A tube containing calcium chloride absorbed the water vapor, and the carbon dioxide then passed through five bulbs filled with an aqueous solution of caustic potash (potassium hydroxide); the gas combined with the potash, forming potassium carbonate. Determining the weight of the calcium chloride tube and the five-bulb apparatus before and after the absorption of water vapor and carbon dioxide allowed Liebig to indirectly measure the mass of hydrogen and carbon in the original sample. If there was any nitrogen in the sample, it could then be volumetrically measured by a separate technique after the carbon dioxide and water had been removed (Rocke 2000b: 282–4).

Although the five-bulb apparatus was challenging to construct, the *Kaliapparat* was safer to use and required less skill to operate than traditional volumetric apparatus. Most importantly, it was accurate. Because large volumes of gas no longer had to be contained, Liebig could burn larger sample sizes at higher temperatures, increasing the likelihood that the combustion reaction would run to completion, i.e. there would be no unburned residual material (Rocke 2000b: 284–5; Usselman et al. 2005; Jackson 2015b).

Complete combustion also reduced potential sources of error. Liebig wrote that "no greater misfortune can befall a chemist than being unable to disengage himself from preconceived ideas [...] If to a beginner in analysis I give a mineral

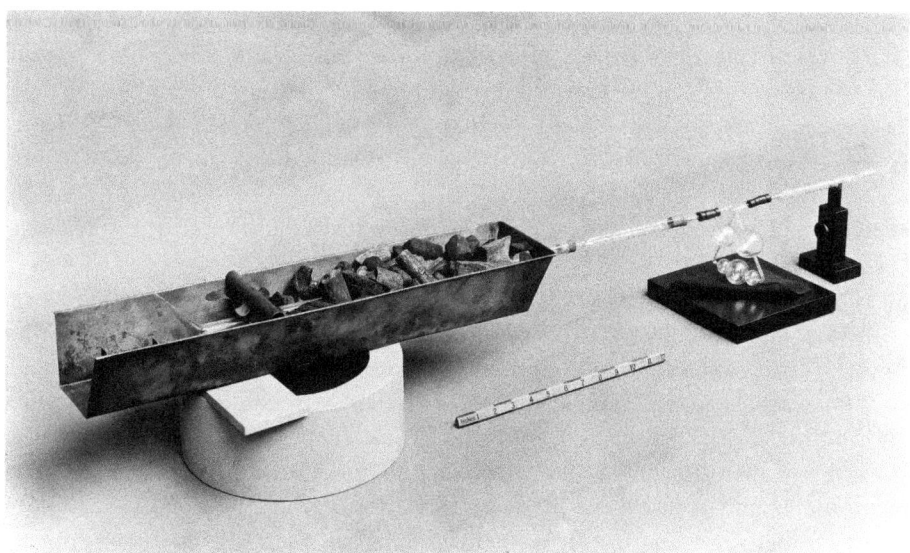

FIGURE 3.6 Liebig's *Kaliapparat*. On the left is a small tray to hold burning coal that would heat the sample in an enclosed glass tube. The water produced during combustion was absorbed by calcium chloride in the tube immediately to the right of the tray. The heart of the apparatus was the five-bulb fixture on the right that captured carbon dioxide with a saturated solution of aqueous potassium hydroxide. Courtesy of the Wellcome Collection.

with the remark that he must look for antimony, lead, and potassium, I am sure he will find antimony, lead, and potassium in spite of all contradictory evidence" (Hofmann 1876: 67). Developing carefully constructed and tested techniques of chemical analysis was key to avoiding experimental bias. The *Kaliapparat* and the spectroscope reduced the ambiguity associated with more qualitative assessments of chemical composition, and established more rigorous and quantitative standards for chemical analysis.

While this equipment facilitated the study of chemical composition, chemists employed other techniques and instruments to probe chemical structure. One particularly important method was polarimetry. In the seventeenth century, Christiaan Huygens discovered that light passing through two crystals of Iceland spar or calcite varied in its intensity when the orientation of the second crystal changed. In the early 1800s, the French engineer Etienne-Louis Malus showed that this phenomenon could be caused not only by doubly refracting materials like spar, but also by reflected light under certain conditions. He referred to this effect as polarization (Mauskopf 1976: 56–8). By studying the refractive properties of a variety of different materials, French natural philosopher Jean-Baptiste Biot, among others, realized that this property might prove useful in chemical analysis. He discovered that the orientation of the plane of polarized light was affected by "certain organic liquids, vapors, and solutions, including oil of turpentine, cane sugar solution (which had a leftward deviation), as well as beet sugar and natural camphor in alcohol, which rotated the plane to the right" (Mauskopf 1976: 62). Biot worked to standardize these observations: "the number of degrees of rotation" of the plane of polarized light could be determined and became "a measure of the 'optical activity' of the substance" (Turner 1983: 222).

The resulting polarimeter embodied decades of experimentation on the nature of light and the refractive properties of matter (Figure 3.7). It consisted of a light source, a liquid-filled tube, and two Nicol prisms, i.e. a prism constructed out of two pieces of Iceland spar glued together with balsam fir resin. The first Nicol prism polarized the light from the lamp. The polarized light then traveled through the solution-filled tube to the second Nicol prism. If the tube was filled with water and the two prisms were at right angles to one another, then no light escaped from the instrument. If the tube was then filled with an optically active solution, then the second prism would have to be rotated either to the left or to the right to block the light (Morris 2015: 163). This angle of rotation could be accurately measured and depended on the chemical composition and concentration of the solution, the color of the light source, and the temperature of the system.

The polarimeter became an instrument of research and industry. Using this instrument, chemists continued to build on Biot's work in several ways. By investigating the optical activity of tartrate salts associated with wine-making, Louis Pasteur discovered the phenomenon now known as enantiomorphism. In the 1840s, he experimentally determined that sodium ammonium paratartrate, an optically inactive material, was composed of two optically active asymmetric crystals that rotated the plane of polarized light with the same magnitude, but in opposite directions (Mauskopf 1976: 72–9; Ramberg 2003: 32–5). As Pasteur described it, "the tartrate salt consisted of 'right-handed' crystals and the paratartrate salt consisted of an equal mixture of 'right-handed' and 'left-handed' crystals," which explained why it had no effect on polarized light (Ramberg 2003: 33–4).

In studying the isomers of lactic acid – only one of which was optically active – Jacobus Henricus van 't Hoff and Joseph Achille Le Bel proposed that optical activity was due to the geometrical position of atoms in the molecule in a manner "entirely analogous to the arrangement of molecules in the optically active crystals ... discovered by Pasteur." In 1874, van 't Hoff theorized that within organic compounds, each carbon atom normally combined with four other groups to form a tetrahedron, but that the "optical activity in a compound [was due] to the presence of at least one asymmetric carbon atom in [the molecule's] structure." He observed that "in several reactions of optically-active

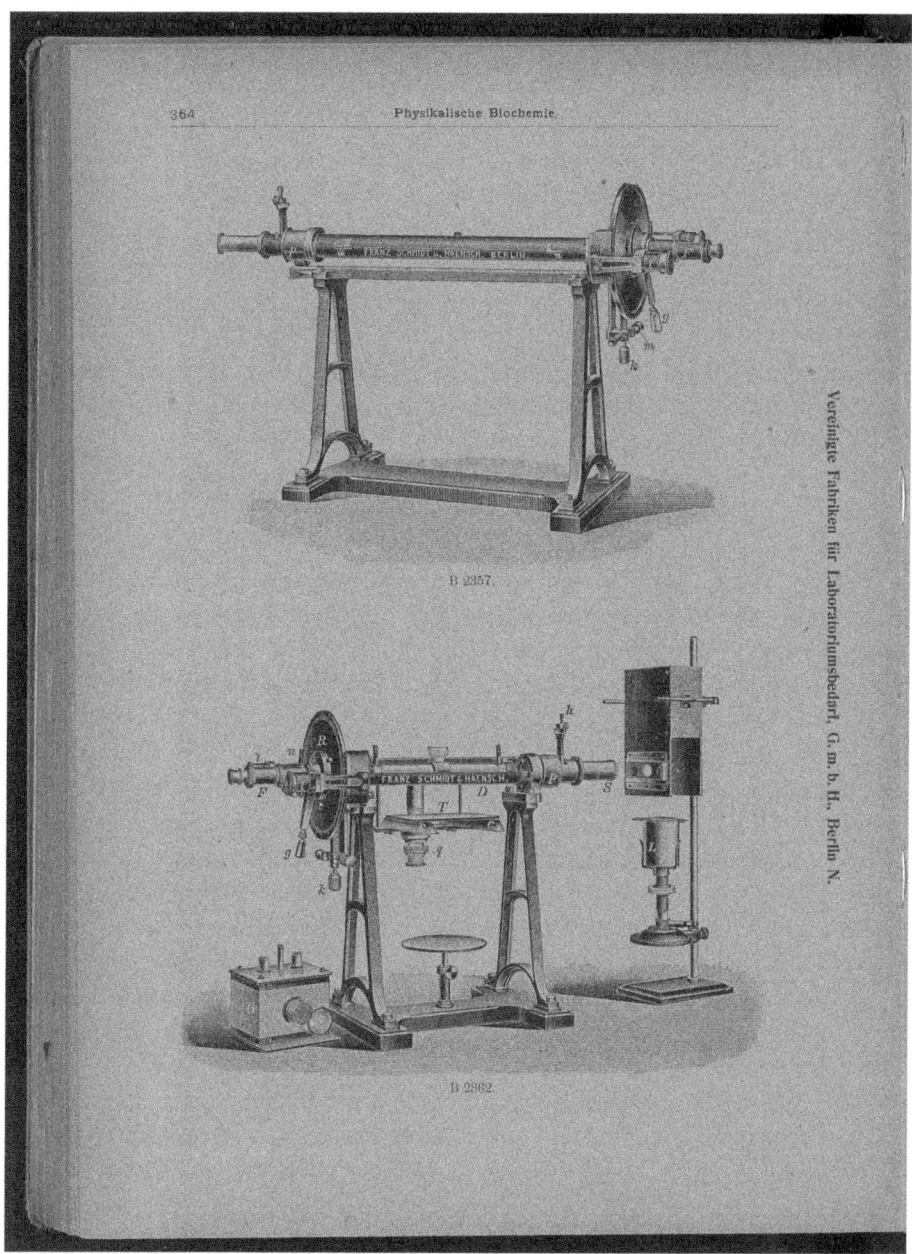
FIGURE 3.7 The 1913 Franz Schmidt & Haensch polarimeter, a modified version of the Jellett–Cornu polarimeter. Courtesy of the Science History Institute, Philadelphia.

compounds, the removal of an asymmetric carbon atom resulted in the loss of optical activity," which supported his hypothesis (Ramberg 2003: 57–9).

Thus, the polarimeter allowed for the chemical identification of similar but subtly different organic compounds. Because sugar was an optically active substance with numerous isomers, polarimeters were widely used in the food industry to monitor the quality of cane and beet sugar used in bakeries and breweries (Jago 1886: 445–8). Additionally, Emil Fischer "realized that van 't Hoff's theory of the asymmetric carbon atom could be [further] employed to classify the increasing number of sugars and their derivatives isolated in the laboratory" (Ramberg 2003: 243). In 1894, for example, Fischer used the polarimeter – coupled with van 't Hoff's theory – to identify sixteen stereoisomers of aldohexose, a carbohydrate containing six carbon atoms and an aldehyde group (Morris 2015: 165).

The polarimeter was just one of many instruments developed by physicists and adopted by chemists to shed light on chemical composition and change. Because of the challenges in using a chemical approach, which often changed the very nature of the chemical reaction under investigation, Ostwald, like van 't Hoff, focused on measuring the physical properties of chemicals in solutions, such as their viscosity, conductivity, and refractive index. His instruments, for example, included the polarimeter and the manometer, the latter of which was invented in the 1600s by Evangelista Torricelli and improved by Emile Amagat in the 1870s, to measure minute pressure changes in containers housing chemical reactions. He also used the electrometer (or galvanometer), developed by Alessandro Volta and modified by Faraday, for electrochemical experiments to measure the electrical potential differences of materials in solution (Ostwald 1912: 458; Guralnick 1979; Pancaldi 2003). For each experiment, Ostwald thought carefully about what each device could measure as well as its limitations; instrumentation was the linchpin of his experimental research program.

Devices invented in the eighteenth century took on new importance in nineteenth-century chemistry. Working with Pierre Simon de Laplace and instrument maker Pierre Bernard Mégnié in the 1780s, Lavoisier created the calorimeter to measure the amount and direction of heat transfer during state changes and chemical reactions (Lavoisier and La Place 1783; Beretta 2014: 207). Building on Lavoisier and Laplace's investigations as well as developments in thermometry, Russian chemist Germain Hess' experiments on the heat evolved in the formation of salts in the 1830s and 1840s showed that, regardless of the number of steps involved in the reaction, the total amount of heat generated between the initial and final stages of the reaction did not appreciably change, which facilitated the development of thermochemistry (Ostwald 1912: 272–6; Leicester 1951). In 1787, Abraham Bennet, a British minister and amateur electrical researcher, invented the gold-leaf electrometer: the relative displacement of its leaves stood as a measure of the strength of weak electrical effects in the atmosphere (Bennet 1787; Elliott 1999). Renamed the electroscope in the late nineteenth century, this instrument became an important tool for Pierre and Marie Curie in their study of ionizing radiation produced by radioactivity (Malley 2011: 23–6).

CONCLUSIONS

While working with wood, brass, rubber, and glass at the bench, chemists saw significant improvements in their laboratory spaces. As the chemical industry prospered, university laboratories benefited from increased state funding as illustrated by the growth in both the physical size of chemical laboratories and the number of enrolled chemistry students. The design and proliferation of specialized rooms and the use of fume hoods and water, gas, and sewer lines also required support at the institutional, local, and state level, which chemists received.

With these improvements, Freer wrote, "chemistry is growing to be more of an exact science every day" (1895: iii). Although chemists continued to use instruments developed in the 1700s such as the calorimeter, thermometer, and distillation apparatus, they also devised new technologies to investigate and deepen their understanding of chemical composition and structure. The spectroscope provided a novel approach for identifying the chemical constituents of materials based on each element's characteristic spectrum. The *Kaliapparat* allowed organic chemists to gravimetrically determine the percentage of carbon, hydrogen, and nitrogen present in complex substances, while the polarimeter shed light on the chemical structure of compounds by measuring their optical activity. The development and modification of chemical technology illustrated chemists' commitment to improving their craft through precision instrumentation and quantitative studies of chemical composition and structure.

CHAPTER FOUR

Culture and Science: *Chemistry Spreads Its Influence*

AGUSTÍ NIETO-GALAN AND PETER J. RAMBERG

INTRODUCTION

The nineteenth century was the "age of science," when natural philosophers became "scientists" within distinct professional disciplines in schools, universities, and industry (Knight 1986). As chemistry established itself as a new discipline, it also crossed numerous intellectual and cultural boundaries, diffusing into the other sciences, into industry, and into the public sphere. Chemical analysis and theory influenced physiology, agriculture, and medicine, and sparked debates on the very nature of life. Chemists also turned their interests toward such physical phenomena as color, light, heat, electricity, and magnetism, influencing physics and astronomy. Chemistry in turn was also influenced by physics and the new science of thermodynamics that provided a method for understanding and quantifying chemical change and equilibrium, enabling chemists partly to accomplish the Newtonian dream of the mathematization of natural phenomena beyond mechanics and gravitation. Chemistry also affected the culture of the chemical industry, transforming it into a science-based industry, a successful combination of pure and applied chemistry, that relied on close collaboration between academic and industrial chemists.

Chemistry moved into the public sphere with popular lectures and books. Humphry Davy and Michael Faraday offered regular lectures on chemical phenomena with spectacular demonstrations for a popular audience in London at the Royal Institution. Justus Liebig's *Chemische Briefe* (1843) described clearly the basics of chemical theory and how it applied to everyday life. Chemists cultivated the idea that chemistry was a natural philosophy with intellectual merit that also had significant economic value to society when its principles were applied to medicine, agriculture, and the production of material goods. Many of these lectures and popular books would also in turn inspire young men, and sometimes women, to take up chemistry as a career.

Chemistry has recently been labeled as the "central science," because it serves as the foundation for many other sciences. As we shall see, the foundations for this central nature of chemistry were formed in the nineteenth century. Nineteenth-century chemists were confident that chemistry, the science of material substances, could be applied in almost every other area of science and to the material production of industry. While they may not always have been successful with their specific ideas, the overall culture of chemical theory and practice had a deep and lasting influence on other sciences.

CHEMISTRY, BIOLOGY, AND MEDICINE

The life and works of the French chemist Louis Pasteur exemplify the spirit of chemistry in the nineteenth century as the "central" science tied to physics on the one hand and biology on the other. Pasteur was trained as a chemist during the 1840s, and studied closely the relationship between crystalline form and chemical composition. The science of crystallography, closely related to geology, had already influenced chemistry, including Eilhard Mitscherlich's theory of isomorphism in 1819 and Auguste Laurent's suggestion during the 1830s of a hypothetical molecular prism as a basic unit to explain substitution reactions in organic chemistry (Melhado 1980).

While working with crystals of tartrate salts, Pasteur noticed they had unusual properties with polarized light. Some tartrates would rotate a beam of polarized light, while others did not, indicating to Pasteur that the molecules of tartrate were asymmetric, and could be either "left" or "right" handed. Living organisms, according to Pasteur, created asymmetric molecules, so life itself must occur by asymmetric processes. This result would prompt Pasteur further into the nature of life, including his more famous research on fermentation, microbiology, and vaccination, and his successful campaign against spontaneous generation – the idea that life cannot emerge from non-living matter (Geison 1995). Pasteur's science had a profound impact on public hygiene, on the medical profession, and even on French colonial interests. He deftly managed to convince farmers, industrialists, politicians, and fellow scientists of the crucial agency of microorganisms in causing disease and alcoholic fermentation. Pasteur's funeral in the streets of Paris in 1895 was a clear sign of his tremendous popularity and the triumph of the so-called "pasteurization" of France (Latour 1993).

Pasteur's own move from crystallography and chemistry to fermentation and microbiology during the 1860s is just one example from the complex history of the interaction between biology and chemistry during the nineteenth century. Pasteur's experiments were aimed at discrediting the idea that life was fundamentally chemical in nature, a position chemists had promoted since the late eighteenth century, based on the successful isolation and characterization of many specific compounds from plant and animal extracts.

As the nineteenth century progressed, chemists had increased the number of these compounds and determined their chemical composition. Michel-Eugène Chevreul had determined that plant and animal fats contained compounds with a specific fixed chemical composition. Urine contained urea and uric acid, two compounds high in nitrogen content. A number of French chemists, including Louis-Nicolas Vauquelin, Pierre-Joseph Pelletier, and Edmond Frémy, submitted brain tissue to chemical analysis, finding numerous fats and cholesterol (Tower 1994). William Prout studied the chemical processes behind digestion, and in 1823 reported the sensational news that the stomach of several mammals contained hydrochloric acid (Coley 1996). In 1834, Prout suggested that foods could be divided into three groups, saccharine (carbohydrates), oleaginous (fats), and albuminous (proteins), each with different characteristic elemental compositions. This threefold division of essential foodstuffs was supported, Prout argued, by the presence of all three in milk, the only natural substance produced as a nutrient (Holmes 1964: liv). In the Netherlands, Gerardus Mulder isolated several substances from both plants and animals that were high in nitrogen content and were very similar in elemental composition. He named these substances "protein, because it is the origin of very different substances and therefore may be regarded as a primary compound" (Fruton 1972: 95–9).

Mulder's results inspired Justus Liebig to have his students study the composition of proteins further, which led him to write one of the most influential books on physiology in the nineteenth century, *Animal Chemistry* (1842). When he heard of Mulder's results, Liebig had already become less interested in the theoretical debates of organic chemistry, in part because he found himself no longer able to lead the field as he once had. He decided to strike out in a different direction and promote chemistry as the fundamental science that would improve our understanding of biological phenomena and enhance commerce (Brock 1997: 92). An inspiration for this move was the encouragement by British colleagues during his first visit to Britain in 1837 that he write up more formally his ideas on the relationship between chemistry and agriculture.

The resulting book, *Chemistry and its Application to Agriculture* (1840), preceded *Animal Chemistry* by two years and became an international bestseller with translations into eight languages. It turned Liebig into an international public figure. Liebig was not the first chemist to offer a scientific approach to improving agriculture (for example, Humphry Davy had already written a well-known book on the subject), but Liebig was able to integrate many earlier ideas about the chemistry of plants with his own innovative ideas: elaborating a nitrogen cycle, explaining the role of minerals in plant metabolism, and refining the law of minima – that plant growth is determined by the necessary element that is present in the least quantity in the soil. The ancient practice of using manure provided plants with necessary minerals in the form of potash, lime, and phosphorous, and Liebig suggested that using inorganic versions of these minerals would increase yields (Rossiter 1975; Brock 1997).

Liebig's suggestions for improving crop yields arrived at an opportune time, as the decade of the 1840s was struck by crop failures and famines, making acute the need for more efficient farming techniques. Although many of Liebig's specific theories were disputed at the time, he nevertheless stimulated research into the role of specific fertilizers on crop production, and created a demand for inorganic fertilizers. Liebig's ideas had a remarkably strong influence in the United States, where he had explained for the first time certain empirical farming practices, and inspired the creation of the first agricultural research stations (Rossiter 1975).

Whereas Liebig's *Agricultural Chemistry* treated the chemistry of plants and its practical application, in *Animal Chemistry* he formulated a thoroughly chemical understanding of nutrition, digestion, disease, and the origins of animal heat. Liebig's fundamental approach involved the "balance sheet" method that had worked well for organic chemistry: the amount of carbon, nitrogen, oxygen, and hydrogen coming into an animal's body must equal the amount leaving it. For example, in November 1840, Liebig performed an experiment with 855 soldiers in the Hessian army. He measured the average amount of carbon the soldiers took in by food, and the amount of carbon they excreted. The difference could then be calculated as the amount of carbon burned daily in respiration (Holmes 1964: xxv). According to Liebig, the source of animal heat, following a tradition begun by Lavoisier, was a slow combustion of carbohydrates and fats to release carbon dioxide. Animal motion, on the other hand, was caused by the decomposition of proteins (high in nitrogen content) in muscle tissue to urea and uric acid, also high in nitrogen, where they were excreted in urine.

While many other physiologists found Liebig's specific theories on plant and animal chemistry wanting (in particular his theory of muscular action), it is difficult to underestimate his influence on the spread of chemical methods and ideas into agriculture, biology, pathology, and physiology. By the time Liebig published his ideas, experimental physiology and organic chemistry had become well established such that Liebig's ideas could be tested and elaborated, rejected, or expanded to a broad range of normal and abnormal physiological processes. Physiologists, especially in France, trained in vivisection and other biological methods, remained skeptical about claims of reducing biology to chemistry. Chemists, on the other hand, were more optimistic: they argued that organisms are complex systems of chemicals whose composition can reveal the nature of physiological processes and the origin of animal heat.

One of the most basic disagreements between biologists and chemists lay in how they explained the nature of alcoholic fermentation. In the early nineteenth century, chemists had largely agreed that alcoholic fermentation was a chemical process, a spontaneous decomposition of the ferment that transmitted its "vibrations" to the sugar molecules to break them down into alcohol. By the 1840s, the issue was no longer so simple, as microscopic studies of yeast indicated that it reproduced, that is, yeast was a living fungus that produced alcohol by physiological, not chemical action. Nevertheless, many chemists continued to advocate for a chemical origin of fermentation. Berzelius, for example, suggested that yeast has a certain "catalytic force" that decomposes sugar to alcohol. Liebig suggested that the ferment was unstable and decomposed, causing another substance in contact with it to decompose (Fruton 1972: 44–7).

Liebig's theory was severely challenged by Pasteur in 1860, who showed that fermentation required living yeast cells. Although the situation was complicated by the isolation of several "soluble ferments" found in saliva and gastric juice, for example, and the broadening of the term fermentation to include many other biological processes besides alcoholic fermentation, there remained a fundamental divide between chemical and biological explanations of fermentation that would not be resolved until the 1890s, with the emerging enzyme theory of fermentation. Many "soluble ferments" had been characterized since the 1840s, but all of them had been isolated from secretions of cells, not in the cells themselves. In 1897, Eduard Buchner showed that juice pressed from ground yeast was active in fermentation, suggesting for the first time that enzymes were responsible for chemical transformations within the cell. Buchner's discovery would eventually resolve the controversy: fermentation is done by chemical agents (enzymes), but those chemical agents are made by living cells. As chemists, physiologists, and microbiologists came to agree after Buchner's discovery, the physiology of cells was mediated by enzymes as chemical agents, and they began to refer to their new hybrid field as "biochemistry" (Kohler 1971; Kohler 1973).

As chemistry made inroads into biological phenomena, it also moved into medicine in numerous ways. New chemicals such as ether and chloroform were used as anesthetics during surgery. In the 1840s, Joseph Lister's successful antiseptic treatment strengthened Victorian values of physical and moral cleanliness (Fox 1988). Experimental physiologists increasingly used chemical tests in their search for new therapeutic substances (Lesch 1984). Although not all physiologists agreed on a physicochemical (reductionist) explanation of life, laboratory medicine and experimental physiology increasingly appropriated the language and tools of the new organic chemistry (Kremer 2009).

By the end of the nineteenth century, clinical chemistry, the use of chemical analysis on bodily fluids for therapeutic purposes, had become a well-established medical field. Its rise can in part be attributed to Liebig, with his enthusiasm for promoting chemical theory as applicable in a wide variety of disciplines and ability to train students in the new methods of organic analysis and place them in prominent faculty positions. In all, thirteen of Liebig's students, trained primarily as chemists, not as physicians, went on to hold faculty positions at medical schools in Germany and Great Britain. In 1842, for example, on Liebig's urging, his student Johann Josef Scherer was appointed Professor of Organic Chemistry at the medical faculty of the University of Würzburg, where he directed the "clinical chemical laboratory" of the Julius Hospital. Liebig's influence was most successful in Germany, where hospitals created clinical laboratories. In Great Britain, clinical chemistry was promoted by Liebig's British students, although it was slower to adopt clinical chemical laboratories. Clinical chemistry made slower inroads in France, where physicians subordinated chemical and physical methods to traditional medical diagnostic procedures (Büttner 2000).

Chemical methods also made their way into forensic medicine, filling a need for expert witnesses in murder trials in the European legal system. The earliest texts on toxicology based on chemistry were written in France by Mateu Orfila (*Traité des poisons*, 1814–5) and in Britain by Robert Christian (*Treatise on Poisons in Relations to Medical Jurisprudence, Physiology, and the Practice of Physic*, 1829). Both books were translated into many languages, and became standard references in forensic chemistry. In the 1820s, Orfila published reliable chemical techniques for definitively identifying blood stains in a murder victim's clothing, and was involved later in controversies over differentiating between human and animal blood by microscopy and smell (Bertomeu-Sánchez 2015).

The most pressing need, however, was to determine the presence of poison (arsenic was the favorite among murderers) in a suspected murder victim's body, because poisoning crimes were otherwise very difficult to detect (Watson 2010: 65). In 1836, James Marsh developed a reliable test for small quantities of arsenic, but it required great skill, took several hours, and was subject to errors and misleading results (Watson 2006: 387). The uncertainties surrounding the Marsh test became evident in the most famous poisoning case of the nineteenth century, the trial of Marie Lafarge, accused of poisoning her husband with arsenic. Local doctors performed repeated Marsh tests for arsenic in her husband's body that gave negative results. The court consulted Orfila, who rejected the results as careless and sloppy. The court eventually asked Orfila to perform the tests himself, and he produced a positive test for arsenic in the victim's organs and Lafarge was convicted. The Lafarge case was known world-wide, and resulted in the close consideration of methods in toxicology, including the reliability and sensitivity of the tests that depended on a high level of skill and sophisticated laboratory conditions (Bertomeu-Sánchez and Nieto-Galan 2006). As a result of the Lafarge case and others like it, courts increasingly relied on experts in toxicology, often drawing them from the pool of academic chemists (Watson 2006).

CHEMISTRY, THEOLOGY, AND IDEOLOGY

As a science of matter, chemistry also became a powerful tool in support of materialism, and the encroachment of chemistry into biology's territory contributed to the complex debates over vitalism in the nineteenth century. Early in the century, chemists had already been convinced that the compounds found in living things were more complex than inorganic compounds, but still followed the laws of composition. Friedrich Wöhler's synthesis of urea from ammonium cyanate in 1828 is often still portrayed by chemists as a death knell for vitalism, but that view crystallized only much later, after Wöhler's death in 1882 as organic chemists, especially in Germany, created a founding myth for their discipline (Ramberg 2000). Marcellin Berthelot's "total synthesis" of organic compounds in the 1860s – making them directly from inorganic raw materials – was done to strengthen and promote his own positivism, materialism, and religious agnosticism (Russell 1987). Berthelot rejected any non-materialist explanation of matter, and employed "equivalents" rather than atomic weights in furtherance of a strict positivistic chemistry based on experimental evidence. He scorned all speculation about the constitution of matter, including atomism, removed any trace of vitalism, and relied on an experimental science with no influence of religion (Fox 2016b).

In Germany, some physiologists favored an even more radical materialism derived from a physico-chemical understanding of physiology. In 1844, Hermann von Helmholtz and Carl Ludwig signed a manifesto in favor of a reductionist physical and chemical explanation of life. Nevertheless, they were more pragmatic than the prominent metaphysical materialists Ludwig Buchner, Carl Vogt, and Jacob Moleschott (Gregory 1977). French physiologists like François Magendie, Claude Bernard, and Charles Richet were agnostic toward the reductionism of either vitalist or materialist/chemical explanations, preferring simply to establish the observable phenomena of physiological processes and the laws that govern them (Coleman 1971: 154–7; Kremer 2009).

A more sophisticated defense of vitalism would come from Pasteur's conclusion that living things create asymmetric molecules. In his famous 1860 lecture "On the Asymmetry of Natural Organic Products," Pasteur declared that the presence of asymmetry was "perhaps the only well marked line of demarcation that we can at present draw between the chemistry of dead nature and the chemistry of living nature" (Geison 1995). According to Pasteur, asymmetric molecules could be produced only by "asymmetric forces," a view closely related both to his conviction that alcoholic fermentation was a vital and not a chemical process, and his later attempts to disprove spontaneous generation (Geison 1995).

The origin of the asymmetry in biological molecules proved difficult to explain in physical or chemical terms. In 1898 Frances Japp argued explicitly that asymmetry at the molecular level required a non-material cause, "a directive force … of precisely the same character as that which enables the intelligent operator, by the exercise of his Will, to select one crystallized enantiomorph and reject its asymmetric opposite" (Palladino 1990). In 1894 Emil Fischer suggested an unequivocally chemical and mechanistic view of fermentation and enzyme action, where the source of biological asymmetry was the asymmetry already present in the enzymes that fit asymmetric molecules like a lock and key. Japp deftly countered Fischer by noting that even if this mechanistic interpretation were true, it still left the origin of asymmetry unexplained.

Reactions to materialism could also be found among chemists who sought a natural theological explanation to reconcile the wonders of the universe with God's original design. Although the positivistic trends of the nineteenth century developed the narrative tradition of a continual "conflict" between science and religion, many natural philosophers attempted to reconcile chemistry as a genuine science of matter with a range of different religious beliefs (Brooke 1991).

William Prout, for example, admired William Paley's natural theology, and in his Bridgewater Treatise, *Chemistry, Meteorology, and the Function of Digestion, Considered with Reference to Natural Theology* (1834), he pointed to Dalton's laws of combination and other recent developments in chemistry as illustrations of divine wisdom. Prout also argued for the specific utility of each of the known elements, and that each newly discovered element would have a distinctive purpose. However, Prout illustrates another, lesser-known aspect of chemico-theology: chemists cannot only imitate nature, but can improve it, working to complete God's creation in a form of process theology. Following this theme, Prout argued that those elements that were clearly dangerous or poisonous could have these effects nullified by turning them into different compounds, and it was within chemists' powers to do so. Chemists acted as mediators, completing God's creation. Prout extended our powers to organic chemistry, and argued that we should be able to make organic compounds, and even improve them (Brooke and Cantor 2000; Brooke 2007). Liebig argued in a similar way that when we understand the divinely ordered laws of nature governing the fertility of plants, we cannot ignore them, and are obligated to use them to our advantage and improve agricultural yields (Sonntag 1977). Chemists would later drop the explicitly theological tone of Prout's and Liebig's view on improving nature, but retain the general notion of improving on nature in a secular context, for example in the burgeoning new artificial dye industry of the late nineteenth century.

CHEMISTRY, PHYSICS, AND ASTRONOMY

On other fronts, chemists interacted significantly with the cultures of physics – especially electricity, magnetism, light, and color. Electrochemistry exemplifies the flexible boundaries between physics and chemistry in the early nineteenth century. Electrical sparks had already been crucial in 1783 for Lavoisier's synthesis of water from hydrogen and oxygen. In 1800, Alessandro Volta discovered the electrochemical battery. In 1804–1805, Humphry Davy's electrochemical experiments led to the isolation of sodium and potassium as new elements, and Berzelius explained the causes of chemical combination through his theory of "electrochemical dualism" – the attraction and repulsion of electropositive and electronegative components of molecules. In the early decades of the century, electrochemistry was practiced under the banner of Romanticism, as in the case of Davy's close contact with poets and writers, but also in accordance with his general interest in the intriguing hidden electrical forces of nature that provided new chemical elements (Cunningham and Jardine 1990). Hidden forces inspired, for instance, Mary Shelley in *Frankenstein, or the Modern Prometheus* (1818), which was superficially a simple fictional horror story, but contained a good deal of scientific (and chemical) knowledge of the time (Shelley 1831).

Liebig's discussion on the origin of animal heat in his *Animal Chemistry* was one of many publications of the 1840s that contained ideas that would lead eventually to the idea of the conservation of energy as the first law of thermodynamics. Liebig's assertion that animal heat and animal motion were the result of chemical transformations fostered in part Hermann Helmholtz's famous 1847 essay on the conservation of force, where Helmholtz wrote that animals consume "a certain quantity of chemical potential energy and produce from it heat and mechanical energy" (Holmes 1964: lxxiii–lxxviii). As physicists developed thermodynamics in the next twenty years, it would in turn work its way back into chemical theory, when late in the century, leading advocates of the new subdiscipline of physical chemistry sought to dig deeper into explanations of the real causes of chemical change. Eighteenth-century qualitative affinity tables and early-nineteenth-century electrochemical explanations were replaced by thermodynamics, which provided tools for the quantification of chemical equilibrium. The famous "ionists," Wilhelm Ostwald, Svante Arrhenius, and

Jacobus Henricus van 't Hoff, based their new experimental and theoretical culture on such physical properties as conductivity, viscosity, osmotic pressure, ionic solutions, catalysis, reaction kinetics, and thermodynamics. In particular, in the 1880s, Ostwald's chemical institute at the University of Leipzig became a research school of international repute and attracted students from all over the world (Servos 1990).

Throughout the century, the borders between chemistry and the other physical sciences would continue to become blurred. The measurement of physical properties played a role in William Ramsay's and Lord Rayleigh's discovery of argon and the other noble gases, in the same way that decades earlier the experimental determination of gas densities had improved the calculation of atomic weights (Ihde 1964; Hirsh 1981; Watson 1995). The electroscope became an effective tool for measuring radioactive decay by the strength of electrical currents, helping to identify new chemical elements by a physical property.

The study of gases would also have implications for meteorology and geology. In 1859, John Tyndall noted that different gases absorb greater amounts of radiant heat, notably carbon dioxide and water. Tyndall was most impressed by water, which he found absorbed eighty times the amount of normal air. For Tyndall, water vapor was the most important gas for controlling the Earth's atmosphere, keeping it warmer than if it was not present. Water vapor, Tyndall wrote, was "a blanket more necessary to the vegetable life on Earth than clothing is to man" (Fleming 1998: 71; Jackson 2018). In 1895, Svante Arrhenius would emphasize the role of carbon dioxide in controlling the Earth's climate as an explanation for the origin of the ice age, by then a well-established geological theory. Arrhenius' model showed that the the reduction of carbon dioxide would decrease global temperature, while increasing it through volcanic activity would warm the Earth's climate. Arrhenius recognized that the increased use of coal in industry would increase the Earth's temperature, but this would be to our advantage, in that "we may hope to enjoy ages with more equitable and better climates, especially as regards the colder regions of the Earth, ages when the Earth will bring forth much more abundant crops than at present, for the benefit of rapidly propagating mankind" (Fleming 1998: 74).

The invention of the spectroscope by the physicist Gustav Kirchhoff and the chemist Robert Bunsen also illustrates how chemistry, physics, and astronomy merged to form the new science of astrophysics. The spectroscope provided a new means of identifying the known chemical elements and for recognizing new elements. Between 1861 and 1863, Kirchhoff studied closely the stellar spectrum, comparing it with the spectra of thirty elements, and identified with certainty six elements in the solar atmosphere. In the early 1860s, William Huggins detected the presence of several terrestrial elements in starlight, and in 1864, announced that nebulae contained prominent lines created by hydrogen.

Huggins' novel method for studying nebula and his results resolved a long standing puzzle among astronomers about the nature of the nebulae. They were not clusters of stars that could not be resolved by the telescope, but as Huggins put it, "enormous masses of luminous gas or vapor" (Longair 2006: 18–19; Becker 2011: 72–6). In 1868, Norman Lockyer discovered a novel line in the solar spectrum that indicated a new element that he tentatively called helium, although he was reluctant to use that name and call it an element until William Ramsay identified it on Earth in 1895 as a product of radioactive decay (Hearnshaw 2010; Becker 2011: 268).

The interaction between chemistry and light was also crucial for photography. Joseph Niépce's heliography appeared around 1822. He coated a sheet of glass or metal with a tar-like form of natural asphalt and exposed the surface to light. After rinsing with oil of lavender, only some asphalt remained, leaving an image on the plate. In 1839, Frenchman Louis Daguerre developed a new and better process that exposed a copper plate covered with silver and vapor iodine crystals to light. This process became known as the "daguerreotype," the first commercial photographic procedure. The same year, Henry Fox Talbot developed a new technique in England, the calotype, which coated paper with sodium chloride and silver nitrate. In the second half of the century, other techniques improved and spread the photographic culture across society. Celluloid, an early artificial polymer formed from plant fibers, grew in popularity during the 1880s and was introduced by George Eastman in 1889 as a flexible backing for photographic negatives. By the 1890s, celluloid was used as film for motion picture cameras and projectors (Friedel 1983).

Astronomers would quickly make use of photography. John Draper made some of the first daguerrotypes of the moon in 1840, and daguerrotypes of solar eclipses (1842), the solar spectrum (1843), and sunspots (1852). The wet collodion process, invented in 1851, shortened the exposure times considerably, but were also prone to evaporation and therefore still limited the exposure times required to image faint astronomical objects. The invention of dry plates in the 1870s allowed for much longer exposure times, allowing fainter stars to appear on the plates. Draper's son Henry would optimize the mechanical and optical equipment necessary for astrophotography in *On the Construction of a Silvered Glass Telescope, Fifteen and a Half Inches in diameter, and its Use in Celestial Photography* (1864), which remained in print for the next fifty years (Lankford 1984; Longair 2006: 7–11).

The relationship between color and chemistry also appeared in a broader commercial and cultural context. Early in the century, Goethe's *Farbenlehre* (1810) had introduced a new way of understanding color from a holistic perspective, which included artisan skills and artistic sensitivity. Goethe's romantic natural philosophy radically opposed Newton's mechanical, analytical study of color, which had begun with his famous experiments on the decomposition of light with glass prisms. Goethe rejected any kind of corpuscular mechanical philosophy that reduced the phenomenon of color to a set of geometric, mathematical simplifications, and defended the romantic sensitivity of painters, artists, and craftsmen as an integral, holistic means of study. Goethe's approach to color influenced calico-printers, dyers, chemists, painters, and even industrial entrepreneurs, who promoted qualitative aesthetic beauty in colored textiles. These artistic and commercial professions contributed as much to the art and theory of color as leading natural philosophers and university professors. Although Goethe's views on color were marginal among natural philosophers, his anti-Newtonian approach opened new frontiers between artisan skills, artistic sensitivity, and academic chemistry (Sepper 1990).

Similarly, the skills required for photography were not far removed from those of dyers and calico printers in fixing their colors to textile fibers and making them resistant to light. The case of the English inventor and calico printer John Mercer is particularly illustrative (Nieto-Galan 1997). Mercer learned some chemistry informally in Lancashire, but acquired solid expertise while printing cotton – calico printing – in several factories, and eventually produced monotone photographs on cotton while experimenting with different colors for textiles. In the 1840s Mercer produced these images by combining metals with natural dyestuffs and inorganic salts and exposed them to sunlight. He presented the results of his photographs at the 1858 meeting of the British Association for the Advancement of Science. Mercer had not only artisanal factory experience, but also benefited from informal meetings with the English chemist Lyon Playfair, a former student of Liebig. Playfair, Mercer, and other calico printers discussed theoretical as well as practical questions regarding the fixation of colors on textile fibers, to the extent that a mere former weaver like Mercer could publish and talk about atomic weights, discover a new method for treating cotton fibers ("mercerization"), and be accepted as a fellow of scientific societies.

Chevreul is another excellent example of the links between chemistry, art, and craftsmanship. From his early research on vegetable fats and his contribution to the emergence of organic chemistry, Chevreul moved towards a more "physical" approach to color. His experience as director of dyeing at the Gobelins dyeworks in Paris – again the artisan connection – meant he was aware of the subtle way in which the customer's eye perceived colors in relation to their neighboring shades, and he developed a law of contrast of colors. Chevreul standardized the colors of the visible spectrum as a circular system, and applied it to many practical cases such as colors on printed paper, textiles, tapestries, and fine-art paintings. His texts and charts had a profound influence on some impressionist and postimpressionist painters and their pointillist style, such as Georges Seurat and Paul Signac. Art historian John Gage described Seurat as a "scientific painter" because of his constant search for a precise and "scientific" harmony of lines, light, and colors. Chevreul's work is still used in art schools today. This is an excellent example of the intersections between chemistry, color, and art (Gage 1993).

CHANGING THE CULTURE OF INDUSTRY

William Perkin's accidental synthesis of mauveine in 1856 created one of the first synthetic dyes that would change chemical industry and have broad cultural effects on fashion. By the 1870s chemists were able to replicate natural dyes and had developed several different classes of artificial dyes in almost any color they wished, and there was high demand for these synthetic dyes in international markets. At first, small dye firms emerged in Britain and France, but in the 1870s, the majority of dye production had shifted to German firms that created a successful new culture of research and development, in laboratories modeled after the many academic teaching–research laboratories at German universities. The academic laboratories would also have a ready supply of new research chemists for industry. Solid academic research training of industrial chemists, and an effective genuine patent system (the German patent act of 1877) that protected the procedure but not the final product, did the rest to strengthen German industrial hegemony.

The conversion to artificial dyes did not happen overnight, however. In the second half of the nineteenth century, a number of dyestuffs – natural and artificial, or chemically modified natural materials – coexisted in workshops, factories, and markets. Some consumers were resistant to these technological changes. In many cases, new artificial dyes were "adopted" by the technological system of natural dyestuffs, in a process of "progressive artificiality," which included natural and artificial materials for decades. In the 1870s, for example Chevreul expressed indifference toward artificial dyes due to their lack of permanence in fibers (Nieto-Galan 2001). Consumers would gradually become accustomed to the new materials and by the early twentieth century artificial materials would became commonplace in everyday products.

The appearance of artificial dyes would also replace the traditional artisanal approach to dying, with its reliance on apprenticeship, tacit knowledge, and oral transmission. The emerging dye industry was created by merging the culture of organic chemistry, strengthened by the new structural theory of the 1860s that created a robust academic research program, and the culture of the chemical industry that relied on the materials produced by this new research program. Chemists in industrial research laboratories harnessed this new chemistry to their own ends, creating a system of "industrialization of invention." At Bayer, for example, the main research laboratory was supported by a chemical library, a control laboratory, and a patent department. The promising research results done on a small scale at the main research laboratory would then be converted to larger-scale production and finally, preparation and promotion of the dyes for sale. Industrial research progressively became more efficient, autonomous, and closely connected with market and consumer expectations (Meyer-Thurow 1982). The end result was a new culture of industrial research and production, and the beginning of the modern science-based chemical industry, using chemical theory to design and create new substances for specific purposes. This process of innovation would subsequently be extended to the large-scale production of drugs, polymers, and other materials in the twentieth century.

All of these new dyes represented a new kind of artificial chemical creativity, because a rainbow of new colors were now available that had never before been seen in nature; this revolutionized fashion and created new consumer demands because it made brightly colored clothing inexpensive and available to nearly everyone (Forster and Christie 2013). The new industry also had a significant economic impact on India by 1900, as artificial production of indigo in German factories began to significantly erode the demand for natural indigo from the British plantations. Although there were attempts to compete with artificial indigo, the natural indigo economy would eventually collapse in the first decades of the twentieth century (Travis 1993; Kumar 2012).

CHEMISTRY IN THE PUBLIC SPHERE

By the beginning of the nineteenth century, chemistry had entered public consciousness, but prior to its universal adoption as a school subject, students would often pursue it through informal means such as popular instructional books, home-based experiments, popular lectures, and social soirées. Many chemists embraced entertainment, wonder, spectacle, and even fun, but with the overarching objective of engaging with chemistry through exploring the nature of everyday objects.

One of the earliest treatments of chemistry for a general audience was Jane Marcet's *Conversations on Chemistry*, first published in 1805. The book was widely read, reprinted, and updated on many occasions (sixteen editions in Britain and twenty-three in the United States). In Britain, *Conversations* was appropriated as a guide to popular lectures on chemistry and natural philosophy, whereas in the USA it became a highly successful elementary chemistry text with added general questions, a glossary of terms, critical comments, and guidelines for experiments (Lindee 1991; Rossotti 2006; Leigh and Rocke 2016).

Born into the wealthy family of the merchant Anthony Francis Haldimand, Jane was educated at home, and when she was fifteen, after her mother's death, she took over the household. In 1799, she married the Swiss–British physician, politician, and chemist Alexander Marcet, and benefited from his broad literary and scientific circles. *Conversations* was inspired by Humphry Davy's lectures at the Royal Institution in London, which Marcet enthusiastically attended. She illustrated the text with her own drawings of instruments, and corresponded with many contemporary chemists. She managed to reach a wide audience, particularly women, and encouraged them to go into the experimental sciences. In *Conversations*, Marcet employed the "familiar format" by reviving an old rhetorical technique of using a written dialogue between a rigorous tutor ("Mrs. B") and two enthusiastic pupils ("Emily" and "Caroline") about the chemical side of many everyday experiences (Keene 2013). Marcet promoted an optimistic image of chemistry as a new science in continuous progress, especially after the great achievements of the recent Lavoisierian revolution, but she retained some aspects of the old traditions and old nomenclature that readers could identify with their own daily experiences. *Conversations* inspired many young readers to study chemistry, including Michael Faraday.

The "familiar format" was an important genre in the popular chemistry of the century, and books for the general reader usually adopted a didactic mode of learning about everyday objects in the household. John Joseph Griffin, with his *Chemical Recreations: A Series of Amusing and Instructional Experiments* (1823), broadened the scope by stressing the central role of practical experiments with everyday objects (Gee and Brock 1991). Albert Bernays also used this approach in his popular book, *Household Chemistry* (1852), which focused on a series of linked experiments or observations to build a structure or hierarchy of chemical knowledge (Keene 2013). *Household Chemistry* went through several editions that gradually transformed it into a school textbook. As a group, these books made a remarkable contribution by using "object lessons" to introduce young people to the delights of chemistry.

Many eminent chemists were originally inspired to study chemistry by these informal experiences. Chemistry professors produced textbooks for their students, but also wrote popular chemistry books as part of what they considered their professional duty to spread modern experimental values beyond classrooms and laboratory benches and into society. In addition, popular chemistry texts enhanced their reputations, and generated additional income and better opportunities to attract public and private funding for their research. Among the professional chemists turned popularizers, one of the most successful was Justus Liebig, whose success in writing books on agriculture and animal chemistry spurred him to write books on applied chemistry for an even broader audience. His *Researches on the chemistry of food* (1847), for example, provided advice on cooking to the general population. His method of producing "Extract of Meat" (*Fleischextrakt*), on which he later based his own private company, was hugely popular in the late 1840s among cookery writers.

However, far and away Liebig's most popular book was *Chemische Briefe* (in English, *Familiar Letters on Chemistry*, 1843) that discussed chemical theory and its application to plant and animal physiology, food production, nutrition, and public health, and abstract topics like scientific methodology and materialism. "Chemistry," Liebig wrote at the beginning, was a "most powerful means towards the attainment of a higher mental cultivation," and organic chemistry would provide "the laws of life, [that underlay] the science of physiology" (Liebig in Brock 1997: 282). Written in simple, elegant language, Liebig's chemical letters became a bestseller and a classic of nineteenth-century German literature. Liebig continued to revise and expand the book in multiple editions until 1865, when it would contain fifty essays on all topics of applied chemistry. It was translated into nine languages, and the first English edition alone sold 60,000 copies (Brock 1997: 275).

In France, Louis Figuier wrote essays for popular scientific journals and in such works as *Les merveilles de la science* (1867) and *Les merveilles de l'industrie* (1873), which told general readers about the progress of the scientific culture of the century. Figuier began with an academic career in chemistry, pharmacy, and medicine and worked with Antoine-Jérôme Balard – the discoverer of bromine – at the Sorbonne. However, he was soured by the ethos of professional science, and after a bitter controversy in 1856 with the prestigious physiologist Claude Bernard, he turned to writing popular science. In tune with its revival in the nineteenth century, Figuier used alchemy as a vehicle for science popularization, and in 1860 he published a history of alchemy from antiquity, mainly based on the history of metallic transmutation and the quest for the philosopher's stone in texts by Nicolas Flamel, Jan Baptist van Helmont, and Robert Boyle. He used the conversion of metals into gold by the philosopher's stone as a dream to stimulate readers' imagination and encourage nineteenth-century chemists to develop new experimental techniques. In spite of the distaste for alchemy – associated with theological and metaphysical knowledge rather than contemporary positivist, "modern" science – Figuier believed that transmutation could still be useful as a lure to pique the public's curiosity about chemistry (Nieto-Galan 2016).

Other chemists wrote histories of chemistry for a larger audience. In Germany, Hermann Kopp published the four-volume *Geschichte der Chemie* between 1843 and 1847, and followed this with revised histories in the 1860s and 1870s (Rocke and Kopp 2012: 7). In France, Adolphe Wurtz completed his *Histoire des doctrines chimique* in 1869. He began the book with the controversial phrase: "Chemistry is a French science, founded by Lavoisier in immortal memory," which was likely intended by Wurtz not as a chauvinistic claim, but as a rhetorical plea to his fellow countrymen that the success of German organic chemistry actually rested on French foundations (Rocke 1994; Rocke 2000a). Marcellin Berthelot also wrote such notable historical works as *Les origines de l'alchimie* (1885), *La révolution chimique: Lavoisier* (1890), and *La chimie au Moyen Age* (1893). He presented alchemy as being full of superstition, and as proof of the victory of modern, positive science over primitive errors, which was in tune with Republican values and the alliance between science and free thought (Fox 2012). Like Berthelot, Ostwald was also interested in the cultural and philosophical aspects of chemistry; he included its history as an intrinsic part of his textbooks, and initiated a long-running series of pamphlets that reprinted classic scientific papers. He also published on natural philosophy, scientific progress, the psychology of the scientist, scientific creativity, and even ethics and sociology.

The border between popularization and formal teaching of chemistry was also flexible and elusive. Throughout the century, many of the works written by chemists, from academic textbooks to popular chemistry texts, were widely circulated and were read and appropriated in very different ways. This was the century of the professionalization of chemistry in universities, technical schools, and industries, so its teaching became an important cultural manifestation that deserves further examination. Beyond the tradition of demonstration experiments, Louis-Jacques Thénard's medical and pharmacy students in Paris discussed the results of chemical analysis together with the lecturer. In the 1830s, in Germany, at the small University of Giessen, Liebig engaged his students directly in specific research problems concerning the analysis of animal and vegetable compounds in the laboratory, blurring the boundaries between chemical teaching and research (Brock 1997). However, although Liebig's chemical training was innovative and distinctive in the 1830s, the increasing institutionalization of chemistry at universities and technical schools provided students and teachers in subsequent decades with new laboratories, lecture halls, and other spaces for its teaching. From Henri Saint-Claire Deville's courses at the *École Normale Supérieure* in Paris (Figure 4.1) to Faraday's public lectures at the Royal Institution in London there was a wide variety of modes of chemistry instruction.

Other chemistry sites combined public lectures with original research, such as the Royal Institution, where Humphry Davy and Michael Faraday were masters of the art. The contemporary movements of natural theology and utilitarianism were not in conflict with their work on the new electrochemistry, the systematic study of metals, or the study of electromagnetism. Faraday's private backstage laboratory was complemented by his entertaining public lectures, with spectacular experiments and chemical reactions in the presence of the upper-class and journalists (Figure 4.2). Through Davy and Faraday, the Royal Institution became famous for its lecture-demonstrations both at formal events aimed at scientists and public lectures that attracted the general public, including women. Faraday's Friday Evening Discourses, his Christmas Lectures for Young People, and his articles in the *Literary Gazette*, *The Athenaeum*, and later the *Proceedings of the Royal Institution* itself, were often combined with doses of natural theology and utilitarianism. Public lectures at the Royal Institution would continue throughout the century by other chemists including John Tyndall, Edward Frankland, and James Dewar.

FIGURE 4.1 Henri Sainte-Claire Deville demonstrating an experiment in the inorganic chemistry laboratory at the *École Normale Supérieure*, Paris, 1878. Reproduction after a painting by L. Lhermitte. Courtesy of the Wellcome Collection.

Faraday would turn his last series of Christmas lectures, given in December 1860 and January 1861, into a very successful popular book, *The Chemical History of a Candle* (1861). In a series of carefully prepared lectures, Faraday used the example of a candle, an object that is familiar to absolutely everyone, to construct a general vision of natural philosophy, explaining the nature of the wax, the flame, the chemical process of combustion, and the analogy of combustion to respiration. "There is not a law under which any part of this universe is governed," Faraday wrote, "which does not come into play and is touched upon in these phenomena. There is no better, there is no more open door by which you can enter into the study of natural philosophy than by considering the physical phenomena of a candle." In the lectures, Faraday also expressed an explicitly theistic view that God has written laws into the universe that we must uncover, although this theistic dimension was less evident in the printed version. Beautifully written and illustrated, *The Chemical History of a Candle* was a bestseller and translated into many languages, and has never been out of print since 1861 (Faraday 2011 [1861]).

FIGURE 4.2 Michael Faraday lecturing at the Royal Institution with an audience of many women and some British royals. Photograph by Time Life Pictures/Mansell/The LIFE Picture Collection via Getty Images.

The Royal Institution's lectures and books on "familiar chemistry" highlighted the benefits of experiments, but simple experiments with readily available household objects, as outlined in the books by Marcet, Bernays, and Griffin, could only go so far without additional chemicals and apparatus. In 1836–1837, chemistry sets called "Youth's Laboratory, or Chemical Amusement Boxes" were being sold by Robert Best Ede, including test tubes, a funnel, watch glasses, a spirit lamp, and litmus paper. Instruction books and chemistry sets were gradually linked together to provide material for more detailed investigations, as with Samuel Parkes' *Elementary Treatise* (1839). Many eminent chemists started out with such sets and were known to spend their pocket money on additional equipment and chemicals to extend the experiments. Later, these chemistry sets also allowed children to replicate experiments done at school, thereby putting them in the teacher's place as participants rather than observers (Keene 2013).

CONCLUSION

Chemistry has always been connected to other disciplines – medicine in particular – but these connections became more numerous and stronger during the nineteenth century, as chemical theory became robust enough for chemists to claim that other sciences should rest on its foundations. For the movement of chemistry into biological disciplines, we cannot overestimate Liebig's role in promoting the application of the principles of organic chemistry to physiology, medicine, pathology, and agriculture. His ability to train chemists and place them in academic positions throughout Europe and North America ensured the growth of chemically based biological sciences, even if they left Liebig's own ideas far behind them. As a science popularizer, Liebig relentlessly promoted the idea that chemistry had the power to improve the lives of everyone. As chemistry moved into biology, it immediately conflicted with the theoretical culture of biologists and the issue of vitalism.

Chemistry's interaction with physics may not be as profound as with the biological sciences, yet Liebig's theory of respiration had an influence on the conservation of energy. Chemistry, in turn, drew from thermodynamics to create a unified general chemistry. Bunsen and Kirchhoff's method of spectral analysis, on the other hand, would profoundly change astronomy, with a new method of chemical analysis that allowed determination of the chemical composition of the sun, solar system, and the stars.

The development of the science of chemistry must also be placed in the broader context of the culture of the nineteenth century. Liebig's *Chemische Briefe* is considered a part of German literature, while Goethe's theory of colors closely associated craftsmanship and artistic skills with chemistry workshops and factories. Chemistry also became an important part of profound philosophical discussions on natural philosophy and natural theology in both private and public spheres. It transformed the culture of the chemical industry, replacing the artisanal-based natural dyestuffs trade with carefully planned research and development of artificial dyes. The increased production of artificial compounds also began to change radically the everyday lives of citizens in modern industrial society. Finally, the dimension of chemistry as spectacle, in public lectures, public experiments, and popular texts, and its presence in popular science journals throughout the century, were other signs of its vitality as part of the nineteenth century public sphere.

In her *Conversations on Chemistry*, Marcet well encapsulated the transformational potential of chemistry, when "Mrs. B" remarks to her pupil "Caroline" that a new science called chemistry could serve a broader purpose:

> You confine the chemist's laboratory to the narrow precincts of the apothecary's and perfumer's shops, whilst it is subservient to an immense variety of other useful purposes. Besides, my dear, chemistry is by no means confined to works of art. Nature also has her laboratory, which is the universe, and there she is incessantly employed in chemical operations. You are surprised, Caroline, but I assure you that the most wonderful and the

most interesting phenomena of nature are almost all of them produced by chemical powers ....

(Marcet 1817: 13)

Chemists certainly shared Marcet's opinion that chemistry served an "immense variety of other useful purposes." Chemistry spread through the culture of the nineteenth century and chemists became – beyond their laboratory benches and classrooms – powerful agents in the age of science.

CHAPTER FIVE

Society and Environment: *Increased Access for Women, Growing Consumerism, and Emerging Regulation*

PETER REED

INTRODUCTION

The period 1815 to 1914 saw remarkable changes in the emerging industrial societies in Europe, the United States, and around the world. These nations benefited from the increased production of commodities, whether for personal and family consumption or to meet the growing demands of international trade. However, the expansion of manufacturing and the increase in urbanization brought damage to the environment and public health problems that national governments and local and regional authorities were forced to address through regulatory legislation. Chemists (alongside many other scientists) took a leading role to mitigate the environmental damage and health problems for the benefit of these industrialized societies. The role and benefits of chemistry only became apparent to most people over the course of the period through wider access to education and better communication. Gradually women gained access to education, and as a consequence began to play a more prominent role in academia and the workplace. But the potential for chemistry to make warfare ever more deadly cast a shadow over industrialized nations in the period leading to 1914.

This chapter has six overarching themes: the widening role of chemistry and chemists in industrialized societies; the changing role of women; the increasing contribution of chemistry to everyday life; the adulteration of food and beverages; the role of chemistry in medical improvements; and the regulation of chemicals that pollute the environment, home, and workplace.

CHEMISTRY AND SOCIETY

The nineteenth century saw steadily increasing recognition of the importance of chemistry and chemically trained professionals in the affairs of industrialized and industrializing nations. This trend was aided by the professionalization of chemists and the formation of specialist organizations. As the nineteenth century progressed and educational opportunities expanded, there was increasing recognition of the benefits that chemical knowledge brought to daily life. Industry and commerce began to make wider use of chemistry and chemists to advance their businesses, and therefore increase the prosperity of the populace. However, there were also drawbacks to the increased commercialization of chemistry, including the adulteration of foods and drugs, the availability of chemicals harmful to health, and the damage done to the environment in all its different forms. Nevertheless, chemists trained in the new analytical chemical techniques (who could determine what a substance contained and in what quantities) were able to detect these abuses, and assist governments to regulate them.

In nineteenth-century Britain scientists (and increasingly chemists) were asked to provide science-based evidence to resolve legal disputes. For instance, the 1820 court case of *Severn, King and Company (sugar bakers) versus the Imperial Insurance Company* to recover losses from a major fire required expert testimony from the chemists Michael Faraday and William Brande (Reed 2014: 24). In the 1830s, numerous cases were concerned with emission of noxious "acid gas" from Leblanc alkali works, while beginning in the 1850s court cases were frequently concerned with patent infringement.

Advances in chemical analysis led to the growth of practicing analytical consultants, who offered their services to an ever-widening range of industries: alkali, railways, metallurgy, explosives, and agriculture. Their services were also important for the civil authorities searching for new supplies of clean water to meet the increasing demand of towns and cities. In Britain, analytical chemists were also engaged by the Customs and the Excise to ensure that appropriate duty was paid on taxable commodities, and, from the 1870s, as public analysts for monitoring adulteration of food, drink, and drugs.

Before the 1870s there was little employment of chemists by industry, because common chemical processes had been modified very little from the trade craft. Any modifications were usually overseen by chemists working as chemical consultants rather than regular employees. Such industries included calico-printing, dyestuffs, gas production, soap-making, brewing, metallurgy, as well as the traditional chemical industries such as alkalis, sulfuric acid, and bleaching powder, along with the pharmaceutical industry (Russell et al. 1977: 31–44, 97–103). As chemists became more widely employed in these different industries in a town or region, they often came together to form local societies for the purpose of fostering a better understanding of chemical advances and creating opportunities for social discourse and fraternity.

Major international exhibitions provided a venue for promoting the chemical industry and its wide range of commodities, and they became increasingly frequent during the nineteenth century. The Great Exhibition of 1851 provided a spectacular presence in London's Hyde Park. The massive glass structure housing the exhibition designed by Joseph Paxton became known as the Crystal Palace (Figure 5.1).[1] The exhibition was conceived as

FIGURE 5.1 The opening of the Great Exhibition on May 1, 1851. Engraved by H. Bibby, 1851. Courtesy of the Wellcome Collection.

a celebration of modern industrial technology and design, and even though it attracted exhibitors from twenty countries the overriding aim was to show the strength and supremacy of British industry. The bright, colorful displays of machinery, working models, and industrial products proved very popular and attracted about three million visitors from Britain and abroad during the five and a half months it was open. Chemical companies promoted their businesses with displays that took advantage of their products' spectacular color, luster, and crystalline appearance. The financial surplus from the exhibition was used to create the Natural History Museum, the Science Museum, and the Victorian and Albert Museum in South Kensington. With the Great Exhibition's success, seventeen others followed in the period to 1914, in Europe, North America, and Australia.

The nineteenth century also saw the development of a range of new classes of high explosives that were used in a range of applications from mining to civil engineering. One of the leading innovators, and the most famous, was Alfred Nobel, whose work consisted of a series of remarkable patents for detonators (1864), and dynamite (1866; Haber 1969: 91). Dynamite was a combination of nitroglycerine and kieselguhr (diatomaceous earth, a porous siliceous powder) and functioned as a stable explosive that did not spontaneously explode. By 1873 Nobel had thirteen factories manufacturing these products on a large scale. In 1887 he developed Ballistite (a smokeless propellant based on nitroglycerine and nitrocellulose in the form of collodion) for artillery. In Britain, Frederick Abel and James Dewar, chemists working on behalf of the War Office, developed a variant of Ballistite that utilized nitrocellulose in the form of guncotton, which they named cordite in their 1889 patent. Nobel challenged this patent in a high-profile court case that began in 1892 and ended in Nobel's defeat in the House of Lords in 1895 (Mauskopf 1999; Reed 2017). On a darker side, the development of such explosives revolutionized warfare. With the declaration of war in Europe in July 1914 and the adversarial alignment of the major industrial powers, there was foreboding about the form warfare might take, and how the application of chemical knowledge might lead to the immense destructive power of high explosives and chemical warfare agents (Werrett 2014).

Alfred Nobel died in 1896 and the provisions of his will led to the establishment of the Nobel Foundation (after the resolution of many legal disputes) and awarding of the annual Nobel Prizes. The first recipient of the prize for chemistry was awarded to the Dutch chemist Jacobus Henricus van 't Hoff in 1901 for his work in physical chemistry and chemical equilibria.

WOMEN AND CHEMISTRY

Women have from the earliest times engaged in chemistry in a variety of ways. During the nineteenth century, the role of women would gradually change, although as a whole, women would not participate relatively equally until the late twentieth century. The gradual changes of the nineteenth century were promoted by better access to all levels of education. Early in the nineteenth century, a few notable women were popularizers of chemistry; by the 1890s women began to undertake original research and make important contributions to chemical science. Nevertheless, women struggled to become members of professional societies and gain employment in industry.

Two prominent early-nineteenth-century popularizers of chemistry were Mary Somerville and Jane Marcet; both were self-educated and kept abreast of current ideas by attending informal London social circles, where their knowledge and understanding was appreciated and their contributions welcome, unlike the formal gatherings of professional scientific organizations from which they were excluded. Mary Somerville's *On the Connection of the Physical Sciences* provides the first non-specialist account of research in the physical sciences, including chemistry (Creese 1991: 277). Jane Marcet's *Conversations on Chemistry* was a "didactic dialogue" between a tutor and a pupil that engaged with familiar objects at home through experiments or excursions (Keene 2013: 55). It proved hugely popular over fifty years in Britain and the United States, running to sixteen British editions between 1805 and 1853 (Creese 1991: 277; Lindee 1991; Keene 2013: 55).

Increased access to higher education for women would occur in the last half of the nineteenth century. In Britain several changes brought wider access to higher education including: the foundation of two women's colleges at the University of Cambridge (Girton in 1869 and Newnham in 1871); access to degrees at the University of London (from 1878); opening of the Mathematical Tripos examination at Cambridge (1881); and improved scientific training at Bedford College (1881; Creese 1991: 278). Gradually a steady stream of outstanding women began to put their mark on original research across chemistry and its associated fields.

Improvement in science education in pre-university schools, especially in girls' schools, was a further reason why these higher education initiatives succeeded in Britain. Particularly outstanding was King Edward VI High School for Girls in Birmingham, which had some of the best science laboratories. This school prepared a continuous stream of outstanding young women scientists who would go on to higher education and then postgraduate research. One of the outstanding students from this school was Ida Smedley, who was home-educated by her mother until she went to the High School for Girls. In 1896 she started her studies at Newnham College, Cambridge, obtaining a first class in part one of the Tripos and a second class in part two. Between 1901 and 1903 she did postgraduate research under Henry Armstrong at the Central Technical College in London, was a demonstrator in chemistry at Newnham College (1903–1904) before taking a research fellowship at the Davy–Faraday

Laboratory at the Royal Institution in London (1904–1906). In 1906 she was appointed an assistant lecturer in chemistry at the University of Manchester, becoming the first woman to hold a faculty appointment. She spent much of the later part of her working life at the Lister Institute of Preventative Medicine, and in 1920 became the first woman to be formally accepted into the Chemical Society of London (Creese 1991: 282–3).

In other countries restrictions were often put in place to discourage women's access to higher education, and authorities (almost exclusively male) were reluctant to change the system to attract more women. In France, girls' schools did not teach Latin, a requirement for the baccalauréate examination that provided access to higher education. Such exclusionary practices were also found in Germany, where the *Abitur* (secondary-school leaving examination, required for university admission) was only offered in boys' schools (Rayner-Canham 1998: 44). In the United States, where there was a more general acceptance of women's access to higher education, the main decision was choosing between a women-only or a co-educational institution. Women-only colleges included Vassar (1865), Wellesley, Smith (both 1875), and Bryn Mawr (1884), while co-educational institutions included Oberlin (1833), Antioch (1850), and Berea (1855). Gradually mainstream universities embraced co-education and by 1870 there were eight (Rayner-Canham 1998: 45–6).

In Russia, Yulia Lermontova's attempts to study chemistry created such an intense struggle with the authorities that she was forced to study in another country. Having been drawn to chemistry, possibly by Mendeleev's outstanding work on the periodic table, Lermontova's wealthy father provided the best home tuition as a substitute for school but her application to the St. Petersburg Agricultural Academy was unsuccessful. She realized it would be necessary to study abroad in a country with less-restrictive educational policies, such as Germany. She enrolled at the University of Heidelberg only to find that women were not allowed to be awarded degrees. She persevered and was subsequently allowed to attend Robert Bunsen's lectures and work in his laboratory. After two years she moved to Berlin to work in August Wilhelm Hofmann's research laboratory, where she completed her first research publication in organic synthesis. Thwarted again in her desire to be awarded a degree, she moved to the University of Göttingen, where in 1874 she earned a doctorate, the first for a woman in Germany. She returned to Russia, working first at the University of Moscow and then St. Petersburg, where she investigated the catalytic cracking of petroleum and developed apparatus for the continuous processing of petroleum as a replacement for batch processing (Rayner-Canham 1998: 63).

The most famous and perhaps the most outstanding woman chemist (and physicist) of the nineteenth century was Marie Skłodowska Curie, who became the first woman to be awarded a Nobel Prize (Figure 5.2). In their most famous work, Marie and her husband Pierre spent months isolating minute quantities of polonium and radium from tons of pitchblende ore using

FIGURE 5.2 Marie Skłodowska Curie in her laboratory. Photography by Time Life Pictures/Mansell/The LIFE Picture Collection via Getty Images.

fractional crystallization and other extraction techniques (Rayner-Canham 1998: 99). Marie received the Nobel Prize for Physics in 1903 with Pierre and Henri Becquerel for their study of radioactivity, and she alone (following Pierre's death in 1906) received the Nobel Prize for Chemistry in 1911 for her discovery of polonium and radium and isolation of radium. Even after these awards Marie still experienced discrimination, but her achievements have made her a role model and exemplar of what women can achieve, not just in chemistry or science but in life generally.

CHEMISTRY IN EVERYDAY LIFE

The diversity and quantity of household commodities, many of them drawing on advances in chemistry, grew at an unprecedented rate during the nineteenth century. A look through a typical middle-class home in Britain would reveal a plethora of chemistry-related products (Flanders 2003). Floors might be covered with carpets and rugs carrying sophisticated designs that benefited from the availability of new synthetic dyes, or covered with linoleum, a remarkably hard-wearing floor covering that could have embossed decorations, invented by Frederick Walton and patented in 1860. The walls of sitting rooms might be covered with brightly colored wallpaper that made use of many of the new dyes. An emerald green dye, Paris Green or Scheele Green, proved especially popular even though it contained arsenic (Mertens 2019).

Other decorative features might include decorative metalwork and glassware. Although objects made of silver were presented as gifts for special occasions such as christenings and weddings, silver-plated tableware such as tea services and tureens were decorative but had practical uses. Silver-plating was an electroplating method of depositing a thin layer of silver on a cheaper metal such as copper or on Britannia metal, an alloy comprising copper, tin, and antimony. The principal manufacturer of silver-plated goods was Elkington & Co. of Birmingham (England) who established an international reputation. Some other objects had chrome plating to enhance their aesthetic appearance and durability.

Taking, collecting, and displaying photographs grew in popularity, aided by advances in chemistry that made photographs easier to take, more resilient to damage from light, and more durable to display. After the early work of the Frenchman Louis Daguerre (the daguerreotype in 1839) and the Englishman Henry Fox Talbot (the calotype in 1841), improvements in photography continued at a steady rate. In 1851 Frederick Scott Archer developed the wet collodion process, in 1864 W.B. Bolton and B.J. Sayce reported on their success with an emulsion gel, and in 1884 George Eastman (Rochester, USA) made use of dry gel on paper or film (celluloid). The public's ability to take their own photographs was made easier in 1888 when Eastman started to market the Kodak camera. Photography reached the mass market in 1901 with Eastman's launch of the Kodak "Brownie" camera. Photographs at this time were monochrome, and a reliable color system remained elusive.

Perhaps the most dramatic and far-reaching change in the home through the period was the change in artificial illumination. Earlier in the century candles were still widely used, made either of tallow or the much more expensive spermaceti (a waxy substance from sperm whales). Unfortunately, the former burnt with an unpleasant smell that was absent with the latter, but the availability of paraffin wax in the mid-1850s through the work of James Young made candles inexpensive and odor-free. Candles were made in different forms and also dyed, although sometimes the dyes contained arsenic. However, the luminosity of candles is limited, hence oil lamps burning kerosene derived from coal and bitumen (trademarked by Abraham Gesner in 1841) were widely adopted. Coal gas produced by the destructive distillation of coal in local gas works was utilized increasingly for outdoor and indoor lighting, as municipal and private gas companies expanded the network of gas supplies (mains) across many large towns. The first gas street lighting was demonstrated in Pall Mall (London) in 1807 and installed for Westminster Bridge in 1813, while in 1816 gas lighting was illuminating streets in Baltimore (USA), and by 1820 streets in Paris. Coal gas also created a welcome brighter illumination at home, making reading and other activities easier. However, gas had an unpleasant smell when burnt due to the presence of sulfur, which was only finally removed in the 1880s. The luminosity of gas lighting was greatly enhanced by the invention in 1891 of the gas mantle by the Austrian chemist Carl von Weisbach.

In the United States other sources of illumination – kerosene and acetylene – relied on major technical innovation in the 1850s and 1860s. Abraham Gesner was responsible for the technical advances leading to the production of coal-oil or kerosene from coal and by the late 1850s had overtaken other similar sources because of its safety, brightness, and cost (Lucier 2008: 153), but kerosene was also the conduit leading to exploitation of another industry that was over the remainder of the century (and into the next) to transform the production of chemicals including kerosene. This was the petroleum industry that became a boom industry from the mid-1860s (Lucier 2008: 234). Acetylene gas lamps were used to illuminate buildings, with the acetylene produced by the controlled dripping of water on calcium carbide, a chemical dependent on cheap hydroelectric power for its manufacture (see Chapter 6). As lighting sources, both kerosene and acetylene depended on the development of suitable lamps. Illumination by electric light gradually replaced gas lighting by the early twentieth century, requiring both an electrical supply network and incandescent lamps. Some of the first towns to have limited electrical supply networks were Godalming (Surrey, England) in 1881, Brighton (England) in 1881, and New York in 1882. These networks expanded rapidly in the last two decades of the century, making connections into homes and provision of indoor lighting possible. Many attempts were made to produce a reliable and long-lasting incandescent lamp, incorporating a filament in an evacuated glass bulb. Two of the principal investigators were Thomas Edison in the United States and Joseph Swan in Britain. Both experimented with different materials for the filament: Edison used carbon, whereas Swan used treated cotton thread. The brighter and more resilient tungsten filament was introduced in 1906.

The heating of homes also improved by the end of the century. Most homes relied on coal for both heat and cooking, but most ranges and hearths were poorly designed and caused smoky and unhealthy living conditions. Campaigns and public exhibitions promoting well-designed and smoke-free appliances were organized from the 1870s and were often led by chemists, such as Alfred Fletcher in Manchester. Coal gas was a cleaner form of energy (particularly after removal of sulfur) and as the supply networks grew so did the option of using gas for heating and cooking in the home. Public displays of gas appliances to promote switching to gas often included cooking demonstrations and food tasting. Innovative architects and landlords even took the drastic step of replacing all their coal hearths with a central heating system using either coal or gas. Boilers were usually installed in the basement and the heated air was allowed to circulate through the house via a series of vents. Such installations greatly improved the heating of the home while substantially reducing the consumption of coal and gas and the pervasive grime.

Several commodities regularly found in the home were potentially dangerous. These included matches (or "friction lights" as they were initially known) and arsenic. Matches were developed as a convenient and controlled method of creating fire. Friction lights, developed by John Walker in 1826, were thin splints of wood with one end dipped in a paste comprising potassium chlorate, antimony sulfide, and gum arabic. The splint was drawn quickly through a piece of sandpaper thereby bringing the chemicals into contact with the air and causing the splint to catch fire (O'Dea 1964). These lights were unreliable and sometimes caused injury. Later, the safety match and the "strike anywhere" match were developed. Matches initially used white phosphorus, but this was poisonous and caused injuries to workers at the match factories. In the 1840s much less hazardous red phosphorus was introduced in the safety match: the red phosphorus was incorporated into the striking (or friction) surface of the box in which the matches were stored, and the match head contained potassium chlorate. However, in 1898 the French chemists Henri Savene and Emile Cahen used the non-poisonous phosphorus sesquisulfide for a "strike anywhere" match that could be struck on any friction surface (Crass 1941a).

Arsenic compounds were not only found in decorative dyes but in many other commodities (including medicines and beer), even the elemental form was found in homes (Whorton 2010). It was sold as white arsenic for removing facial hair and for killing vermin such as rats in domestic homes (Picard 2005). Rather than being locked away safely, arsenic was often just left unmarked and so easily mistaken for non-toxic white powders (like baking powder), leading to many accidental deaths, but arsenic also became a poison of choice for those with more sinister motives (Whorton 2010: viii). Lengthy examination of evidence in murder cases focused on whether arsenic was the cause of death, but in 1836 the chemist James Marsh discovered a reliable test that detected even a fifth of a milligram of arsenic (Marsh 1836: 235). The test removed any doubt about whether arsenic was the cause of death.

Other commodities likely to be found in the home might include waterproof clothing, condoms, medicines (patent and prescription), and toothpaste. The Macintosh raincoat was named after Charles Macintosh of Glasgow who in 1823 had registered a patent for a waterproof material produced by sandwiching a solution of rubber in turpentine between two pieces of fabric. The fabric had a tendency to smell and melt in warm weather but a method of vulcanizing the rubber with sulfur developed by Thomas Hancock of Manchester in 1843 helped Macintosh to improve the fabric (Schurer 1951; Raitt 1966: 73).[2] Vulcanized rubber helped the fight against widespread venereal disease (both syphilis and gonorrhea) with the availability of crepe rubber condoms from the 1870s (Lane 2001: 38). The battle to control syphilis was also advanced by Paul Ehrlich's discovery in 1909 of arsphenamine (Salvarsan 606) that was more effective than the widely used mercurial treatment (Figure 5.3). The "606" denoted that

FIGURE 5.3 Salvarsan treatment kit for syphilis, Germany, 1909–1912. Photograph by Science and Society Picture Library/Getty Images.

it was the sixth in the sixth group of compounds Ehrlich synthesized and tested against syphilis. The use of Salvarsan 606 laid the foundation of chemotherapy in the fight against many diseases, especially cancers.

A wide variety of medicines, both patent and prescribed, were found in most homes (Richards 1990; Lane 2001). One of the most prevalent was the reddish-brown bitter tincture of laudanum, a mixture of opium in alcohol in various concentrations. Laudanum was quite inexpensive (cheaper than gin) and therefore readily available. It was used to treat most of the common day-to-day ailments such as colds, coughs, rheumatism, and diarrhea, as well as for treating general melancholia. Even young children were given it on a regular basis. Because it contained morphine it was highly addictive, and its heavy use resulted in many suicides and accidental deaths. In the United States, the 1860s showed that of 200 suicide attempts and accidental deaths, 138 involved laudanum (Calkins 1871: 46). Use was not confined to the lower and middle classes because laudanum was also used by royalty and the landed gentry. Many poets and writers attributed their creative work to laudanum (Hodgson 2001: 48–63).

Toothpowders using chalk and later a paste of hydrogen peroxide and baking powder were popular during the nineteenth century (Picard 2005). The collapsible lead tube for dispensing toothpaste was introduced in 1880, while the first toothpaste containing a disinfectant was marketed in the United States as Kolynos in 1908.

As food and beverages became increasingly commercialized in the nineteenth century, originating from anywhere in the world, adulteration of these commodities became a significant problem. Unscrupulous merchants took the opportunity to alter their product's purity, enhance its commercial value, increase its marketability or even to reduce the excise duty, often using a chemical substance. The proliferation of adulteration was countered by increased attention by authorities to detect adulterated products that took advantage of better and more sensitive techniques in chemical and microscopic analysis (Bigelow 1898: 508). Common commodities investigated included tea, coffee, chicory, bread, alcohol, tobacco, sugar, soap, pepper, beer, and confectionaries.

In Britain adulterations were investigated by the Customs and the Excise and by those concerned about the purity of commodities and any associated health risks. In 1820 the German–British chemist Friedrich Accum published his *Treatise on the Adulterations of Food and Culinary Poisons* that brought to public attention full details of the adulterants while also identifying the offending vendors. However, Accum's exposé did little to stem the practice of adulteration, and it was not until 1850 that events moved towards parliamentary legislation, when Arthur Hill Hassall, a London physician and microscopist, read a paper on "Adulteration of Coffee" to a meeting of the Botanical Society of London. A report of the lecture in *The Times* drew the attention of Thomas Wakley, medical journalist and Member of Parliament who in 1823 had founded the weekly medical journal, *The Lancet*, of which he was also editor. Wakley and Hassall agreed that the latter would write a series of articles based on the analysis of samples of food, beverages, and drugs from shops in and around London in a section of *The Lancet* titled the "Lancet Analytical Sanitary Commission" and included the names and addresses of vendors whose wares were tested. Wakley agreed to bear all expenses and legal liabilities (Clayton 1908: 12). To demonstrate widespread adulteration across the country, Hassall collaborated with other analytical chemists, among them Sheridan Muspratt who had studied under Justus Liebig in Giessen and had founded the Liverpool College of Chemistry in 1848. Muspratt described himself as a "zealous devotee in this cause" (Muspratt 1860: 385).

In 1855 Hassall published *Food and its adulterations; comprising the Reports of the Analytical Sanitary Commission of "The Lancet", for the years 1851 to 1854 inclusive, revised and extended*. It was the most comprehensive study of adulteration to date, and its publication was probably responsible for the June 1855 Adulteration Select Committee. Hassall's evidence to the Committee over several days "proved a very thorough and damning testimony," revealing the wide occurrence of adulterations in common commodities, particularly with tea, coffee, and bread that formed an important part of daily life (see Table 5.1; Reed 2015: 163).

The Committee also sought a working definition of adulteration to differentiate deliberate alteration from the presence of common impurities or accidental contamination.

All tea was usually imported in its natural state, but was often adulterated before the consumer purchased it. Even though the Board of Inland Revenue was examining tea (and many other commodities) by 1844, their concern was to ensure the maximum duty was paid rather than the health of those drinking the tea. Hassall had found multiple adulterants, many thought to be harmful and dangerous to health (see Table 5.1). Coffee was also adulterated with a variety of additives including chicory (see Table 5.1). This proved controversial, because many coffee drinkers preferred some chicory in their coffee. Up to 1840, the sale of mixtures of coffee and chicory was illegal in Britain, but from 1840 these mixtures were permitted, although often with a reduced proportion of coffee because chicory carried less duty. With analytical chemistry and microscopic examination able to differentiate between the two and medical opinion confirming the safety of mixtures, from February 1853 the sale of mixtures of coffee and chicory was allowed only if they were labeled as mixtures; if pure coffee was requested then it had to be supplied (Hammond and Egan 1992: 40–2). To combat the loss of duty the government gradually increased the duty on chicory so it was equal to that for

TABLE 5.1 Common Adulterants from Hassall's Evidence to 1855 Select Committee

Commodity	Common adulterants
Food	
Bread	Alum; lime water; copper sulfate; mashed potatoes
Butter	Water
Confectionery	Compounds of lead, copper and arsenic
Curry powder	Red lead
Flour	Bean flour; rice flour; calcium sulfate
Lard	Alum; potash; potato flour
Pepper (Cayenne)	Red lead
Vinegar	Sulfuric acid
Drinks	
Beer	Lead acetate; arsenic; cocculus indicus
Coffee	Chicory
Gin	Sulfuric acid; water; cayenne pepper
Porter and Stout	Water
Tea	Prussian blue; turmeric; capers; black lead; old tea leaves; plum leaves; sycamore leaves; iron sulfate
Drugs	
Aromatic confectionery (for diarrhœa)	Turmeric; chalk; cassia
Colocynth (purgative)	Wheat flour; seeds of colocynth; chalk
Liquorice	Starch and various flours; cane sugar; gum; chalk; copper; turmeric
Powdered rhubarb	Turmeric; wheat flour

coffee, and as a result little sampling of coffee was undertaken by the Inland Revenue Board after 1863.

Bread was a staple food in most British households and its adulteration was widespread and pernicious (see Table 5.1). Adulterants came from two sources, the flour miller and the baker. The former would often add "moldy flour, the flour of rye, barley, oats, beans, and rice and sometimes even potato in a boiled or crushed state" (Muspratt 1860: 389). To increase the bread's weight and enhance its white appearance, bakers would add alum, blue vitriol, chalk, bone earth, or plaster of Paris (Reed 2015: 162). The use of alum was especially pervasive and caused the bread to collapse into a soggy mass after a day or so that made it unfit to eat. In France and Germany blue vitriol (copper sulfate) was widely used to boost bread's whiteness even though it was illegal.

The report of the Adulteration Select Committee, published in August 1856, provided little legislative direction. Its conclusions highlighted concerns over the fraud of selling products that were not what they were purported to be, but assumed that the public could distinguish between a genuine product and an adulterated one and would be happy to pay for the latter at a lower price (1856). Following a mass poisoning in which arsenic was used in place of plaster of Paris in a popular confectionary, a bill came before Parliament in 1858, but it was not until 1860 that the Act for Preventing Adulteration in Food and Drink was approved. Its terms proved largely ineffectual: the addition of an ingredient that might be harmful to health was outlawed, but no standards were set and the appointment of public analysts was authorized, but not as a legal requirement. A second Select Committee in 1872 was inevitable, and among its conclusions was a suggestion that public analysts (a varied group of practitioners) should meet to discuss the Select Committee's deliberations. The meeting took place on August 7, 1874 and led to the formation of the Society of Public Analysts (Russell et al. 1977: 105–6). In 1875 the much improved Sale of Food and Drugs Act was passed (Richards 1890: 102). Between 1901 and 1914 government regulation ensured minimum standards for food (Rowlinson 1982: 71).

France was even earlier than Great Britain to address adulterations in food. In 1803 the *Conseil de Salubrité* was established in Paris. Similar committees were subsequently set up in other French regions and cities. Their work was assisted by the advances in analytical chemistry made by French chemists. Adulteration of coffee was an early investigation, finding varying amounts of chicory that gave coffee a distinctive taste.

Most other countries drew on the legal decisions in Britain for their legislation to save "much lost time and misspent effort" (Bigelow 1898: 512). In the United States, legislation followed a progressive path much like Britain. Congress passed an act in 1848 to ensure the purity of imported drugs (mainly opium and cinchona bark), a tea adulteration act in 1883, and a law differentiating between butter and margarine products in 1886 (Richards 1890: 102). These piecemeal efforts formed the path to the Pure Food and Drugs Act of 1906, with Harvey Wiley, a chemist and physician who became head of the division of chemistry in the Department of Agriculture in 1883, leading the fight for pure food and the control of adulterations (Young 1989: 4–5). However, individual states also pursued their own legislation. In 1877 several boards of health in New York, New Jersey, Massachusetts, and Michigan adopted laws based on English legislation, but with each state varying its degree of enforcement (Richards 1890: 102).

CHEMISTRY'S ROLE IN MEDICAL IMPROVEMENTS

Besides the health benefits in the home, chemistry contributed to advances in medicine that improved the general health of society, bringing about a lower death rate for children and an increase in life expectancy generally. Advances by Louis Pasteur in identifying the role of germs in diseases, Robert Koch in attributing specific germs to particular diseases, and Paul Ehrlich in pioneering chemotherapy all relied on a better understanding of chemistry. Surgical operations became freer of sepsis and were carried out increasingly without acute pain as anesthesia was introduced. The occurrence of venereal diseases was reduced by both the availability of better condoms and the introduction of Salvarsan 606. Both patent medicines and prescribed medicines found their way in ever larger quantities into households, a demand driven by "spectacular" and colorful advertisements that all too frequently claimed overly optimistic benefits.

Recovery from open surgical wounds was precarious because of their vulnerability to infection from germs in the air or from contaminated instruments. Infections such as septicemia, erysipelas, and gangrene were likely to occur, often resulting in the patient's death. In 1865 Joseph Lister, a surgeon at Glasgow Infirmary, became aware of the use of a disinfectant powder containing phenol (or carbolic acid as it was better known at the time) to treat sewage in the town of Leek (Staffordshire, England) and later in Carlisle (Cumbria, England) with varying degrees of success (Reed 2014: 68–9). After further investigations, Lister used carbolic acid solution in the form of a spray as an antiseptic to treat wounds, and found considerable success in the prevention of post-surgical infections (Figure 5.4).

Advances in anesthesia came about more serendipitously. The young Humphry Davy had studied nitrous oxide at the Medical Pneumatic Institution in Bristol and had then demonstrated its "laughing gas" property during public soirées at The Royal Institution in London when participants experienced a "loss of sensation" while under the influence of the gas. This property attracted the attention of doctors and dentists as a possible means of carrying out procedures without the patient experiencing any pain. The first recorded anesthetic use of nitrous oxide was in 1845 when Horace Wells, an American dentist, successfully extracted several teeth. This led to the discovery of other chemicals having anesthetic properties: in 1846 Dr. William Morton used ether as an anesthetic when removing a neck tumor at Massachusetts General Hospital in Boston, and in 1847 Sir James Simpson, professor of midwifery in Edinburgh (Scotland), introduced chloroform. While there were always concerns that patients might recover consciousness while the operation was still in progress, these early successes prompted the search for even more effective anesthetics.

FIGURE 5.4 Carbolic steam spray made by D. Marr and used by Joseph Lister (1827–1912). Photograph by Science and Society Picture Library/Getty Images.

In addition to general anesthetics, various new local anesthetics were developed. While investigating an improved method for purifying the alkaloid cocaine found in South America coca leaves in 1860, Albert Niemann at the University of Göttingen noted the numbing effect when cocaine was placed on the tongue. This lead to cocaine's wide use as a local anesthetic, even though it is strongly addictive. Pain relief was also aided by the availability of aspirin from 1899, as a synthetic derivative of salicylic acid, found in willow bark (Lane 2001: 165).

CHEMISTRY AND THE ENVIRONMENT

Increasing industrial production and mass migration to cities for factory work created major tensions between the benefits of chemical production and the environment, both large scale (air, water, and land), or small scale (household or working environment). Civic authorities struggled to provide quality housing as well as healthy water and air. Clean water was compromised by contamination from human sewage that was allowed to run off untreated into adjoining waterways. The air of towns was polluted because of the use of coal to generate power for industry and heat homes, while industry released noxious vapors into the atmosphere and untreated liquid effluents into waterways, and dumped solid waste on land adjoining factory sites.

Regulatory action by governments through the second half of the nineteenth century began to address the worst industrial excesses, but always with a fine balance between allowing industry to function without overdue intervention and preventing extreme damage to the environment. Chemists played a leading role enforcing the regulations and monitoring their strict adherence. Health concerns arising from industrial processes were largely ignored until towards the end of the century. As with other public policy legislation, an incremental approach placed increasingly sophisticated chemical products and processes under review while enhancing inspection procedures and measures.

The accelerated pace of industrialization was accompanied by a sharp rise in the consumption of coal by large factories, small businesses (like bakeries), and individual homes, leading to a degradation in air quality (Flick 1980). The air filled with smoke and also oily carbon residues that settled on every outdoor and indoor surface. Trained chemists understood that smoke could be reduced quite substantially by more efficient burning of coal in well-designed furnaces and hearths, and they became leading advocates for effective regulation, although civic authorities were often reluctant to act because of the huge scale of coal burning and the prospective cost of improvements.

In Britain, sporadic attempts to control coal smoke since the eleventh and twelfth centuries when coal was first used as a substitute for depleted availability of firewood had proved ineffectual. In 1853 Lord Palmerston, the former Foreign Secretary who had moved to the Home Office, secured Parliament's approval for The Smoke Nuisance (Metropolis) Act to control black smoke in London. However, after some early successful convictions, enforcement lost its momentum. Many of these attempts to control black smoke through local authorities proved ineffectual, because many of the elected members were either responsible for the pollution or were concerned about the impact of controls on their prospects for re-election.

As one of the prime centers of Britain's mass industrialization, Manchester was especially affected by the effects of black smoke, and by the late 1850s was consuming about two million tons of coal annually (Smith 1979: 199).[3] Robert Angus Smith, a consultant and analytical chemist, experienced Manchester black smoke at first hand. Smith had trained with Justus Liebig in Giessen before spending the rest of his life in Manchester. He used the town as one of his "laboratories" and became the leading international authority on air quality. Many in Manchester drew attention to the deteriorating condition of stonework, bricks, and mortar of the town's buildings. Smith, suspecting coal smoke, analyzed seventy-one specimens of coal and found that some had "as much as five percent or even six percent of sulfur, although the average for specimens used in Manchester was about 1.4 percent" (Reed 2014). Further investigations of the sulfur acids formed when coal was burned led Smith to use the term "acid rain" for the first time in his article "On the Air of Towns" in the *Quarterly Journal of the Chemical Society* in 1859. Even with this better understanding of black smoke and its acid content, Parliament remained reluctant to approve any restrictions on the burning of coal through the period, although some civic authorities made half-hearted attempts.

In the United States many major industrial cities also suffered the adverse effects of black smoke, although in the West concerns focused more on sulfur and arsenic pollution attributed to copper smelters. Major cities affected by black smoke included Pittsburgh, Baltimore, Chicago, Cincinnati, and St. Louis. As in Britain, complaints about smoke were numerous and vociferous, and each city spawned a number of smoke abatement organizations; many such organizations had men- or women-only memberships, while a few were mixed. In Pittsburgh, these included the Women's Health Protection Association of Allegheny County (*ca.* 1890–1900), the Engineers' Society of Western Pennsylvania (1892), the Civic Club of Allegheny County (1897–1919), the Chamber of Commerce (1899–), and the Smoke and Dust Abatement League (1912–) (Uekoetter 2009: 22). Each organization adopted its own specific strategy, from surveillance and inspection, to prosecution or promotion of further electrification. Under the barrage of protests, city authorities approved regulatory laws that amounted to no more than general prohibition orders, because when establishing fines they failed to specify precise levels of smoke. Responsibility for enforcement was assigned to the appropriate city agency; in Pittsburgh this was the municipal works department. If the agency was underfunded and poorly staffed, smoke abatement had a low priority and was largely neglected. Chicago appointed the first smoke inspector in 1907 whose duties included not only inspection and prosecutions but also advising the business community (Uekoetter 2009: 31–5).

In contrast to the United States, Germany had few smoke abatement organizations, inspectors, and regulations on smoke levels. There were many complaints but no organizations with a collective voice that might convince city officials to adopt regulations. On the whole there was little interest on the part of authorities about the issue, although there were exceptions, such as Hamburg's Society of Fuel Economy and Smoke Abatement. From 1910 the academic journal *Rauch und Staub* (*Smoke and Dust*) was published with the aim of "providing a meeting point for the movement against smoke" (Uekoetter 2009: 43). When municipal authorities approved regulations at all, they comprised general statements without sufficient detail for effective enforcement. The primary role of officials was to respond to complaints while avoiding any controversy.

In addition to coal smoke, beginning in the 1820s the air in Britain was further polluted by nuisance vapors from chemical works, in particular from alkali works using the Leblanc process (Barker et al. 1956). Nicolas Leblanc had developed the process in France in the late 1770s for the production of soda (sodium carbonate) from salt (sodium chloride) rather than relying on barilla (a Mediterranean saltmarsh plant) and kelp, whose supplies were affected by European wars. The process was first industrialized in Britain about 1814 and the level of production grew steadily during the remainder of the nineteenth century in response to domestic and export demands for soda. By the 1860s the Leblanc alkali industry in Britain had become a major contributor to the national economy: consuming 1.8 million tons of raw materials, employing 19,000 workers, producing 3.8 million tons of waste materials, and involving £2 million working capital. The quantity of salt consumed had reached about 250,000 tons annually.

The first stage of the process liberated corrosive hydrogen chloride gas that was released into the atmosphere in mounting qualities from ever taller chimneys. The tall chimneys reduced the impact on the immediate area, but also dispersed the acidic gas over larger areas, and was hazardous to people and the natural environment (Dingle 1982). Legal claims against manufacturers proved difficult and costly, and it was only when wealthy landowners saw their property values decline that parliamentary action became inevitable. The leading proponent was Lord Derby, a former prime minister whose large estate at Knowsley Hall outside Liverpool had suffered the effects of acid gas, and in 1863 he persuaded the House of Lords to establish the Select Committee on Injury from Noxious Vapours (or "Lord Derby's Smell Committee," as *Punch* called it). By the 1860s about 155,000 tons of acid gas were produced by the Leblanc alkali works.

Among many witnesses, the Select Committee took evidence from William Gossage, an alkali manufacturer in Worcestershire, who described how instead of releasing the gas into the atmosphere he had dissolved almost all of the gas in water using a derelict windmill tower. The windmill was packed with gorse, brushwood and coke, and then a stream of water together with the acid gas entered at the top. Nearly all the acid gas was dissolved by the water during its passage to the bottom of the mill (Reed 2015). The "acid tower" was a remarkable innovation and although patented in 1836 was not widely adopted until the early 1850s, because alkali manufacturers doubted its effectiveness. There was an assumption that absorbing large volumes of gas required large volumes of water, whereas what mattered was the large surface area of contact between the acid gas and the water as the latter passed over the gorse, brushwood, and coke (Reed 2008: 109). An acid tower of the early 1860s is shown in Figure 5.5. The "acid tower" was not the final solution because there was little use for the corrosive hydrochloric acid produced in the "acid tower" and it was drained into adjoining canals, rivers, and streams. Only in the 1870s

FIGURE 5.5 Acid tower at Messrs. John Hutchinson & Co., Widnes, in 1860 (Lunge 1886). Courtesy of Peter Reed.

when esparto grass was imported for papermaking was the acid converted into chlorine and then bleaching powder.

Convinced of the efficacy of the "acid tower," the Select Committee recommended legislation forcing manufacturers to release no more than 5 percent of their acid gas. Following publication of the Select Committee's report, parliament approved the Alkali Act 1863 to become effective from January 1, 1864. The terms were clearly defined: all alkali works must be registered; at least 95 percent of the acid gas was to be condensed; an inspector would be appointed who would report annually to parliament; and civil action brought by the inspector in the County Court would recover the damages.

The first inspector was Robert Angus Smith, a Manchester-based consultant and analytical chemist (Reed 2014: 110–12). To ensure inspection of alkali works across the whole of Britain, four subinspectors were appointed, each covering a different region (MacLeod 1965). This team formed the alkali inspectorate. With no precedent and in anticipation of giving evidence in a court of law, Smith was determined to rely on analytical chemistry to underpin the inspection procedures. Manufacturers were concerned about inspection of their works but also about the commercial sensitivity of any information gathered by the Inspectorate. To ease this issue, each alkali works was assigned a number. When works were referred to in annual reports they were identified by their number and not by the company's name. Inspection of works was initially a sensitive issue for manufacturers, because it changed the government's relationship with industry from laissez-faire to interventionist. However, the inspector's role as a "peripatetic consultant," working with manufacturers to improve the operation of their processes, gradually allayed their concerns.

The dry copper industry also raised major pollution concerns. Largely confined to Swansea (south Wales) and St. Helens (northwest England), smelting copper was associated with "pernicious and damaging pollution from 'copper smoke' constituting acid rain and fine particulates" (Reed 2012). The Cornish copper ore being smelted in the 1860s had a high content of fluorspar and sulfur, and so created additional "acid rain" of sulfurous, sulfuric, and hydrofluoric acids (Newell 1997). Also, as a result of the process, copper, sulfur, arsenic, antimony, and silver were deposited as particulates over the surrounding countryside. The local population had a fickle relationship with both the industry and the owners of the businesses because the industry had a firm presence as an employer in the area for over a century. There were few legal challenges and the smoke was just accepted as an inconvenience (Newell 1990; Newell 1997). Even the Medical Officers of Health played down the health dangers, although the smelters tried with some degree of success to improve the furnaces and so ameliorate the worst aspects of the process.

Later, further Select Committees were set up to review the alkali inspectorate's work and to consider whether other processes and nuisances should be put under regulation. Public campaign groups such as the Lancashire and Cheshire Association for Controlling the Escape of Noxious Vapours and Fluids from Manufactories (formed in 1874) drew attention to other damaging pollutants such as "galligu" (from Leblanc works), as well as noxious by-products of cement works, potteries, ammonia and fertilizer works, and manure works (Wilmot 1998; Reed 2014: 135–47). The people of Widnes, a major center for the alkali industry in northwest England, conceived the word "galligu" for the black, gooey, viscous, smelly mass of sulfurous solid waste generated from the Leblanc process. Because there was no really effective process for extracting the sulfur, it was dumped on adjoining land or out at sea. Every ton of soda was accompanied by 175 tons of "galligu," which meant for the 1860s about 352,000 tons of waste annually.

The 1872 Public Health Act transferred responsibility for the inspectorate from the Board of Trade to the Local Government Board that also had responsibility for public health. This change raised the prospect of a more focused approach linking industrial nuisances with health, but progress was slow. Several attempts were made by the Chief Inspector over many years to have coal smoke placed under regulation, but there was no parliamentary appetite for such legislation. Nevertheless, although eighty alkali works were registered in 1864 (the first year of the inspectorate), by 1894 there were 1,173, demonstrating the commitment and success of the inspectorate, but also reflecting the expanded list of processes under regulation.

Several works were using the Leblanc process in France in the 1800s, so how did the French tackle air pollution (Smith 1979: 244–61)? By 1815 the location of chemical works was subject to a decree issued by Napoleon in 1810 (Le Roux and Fressoz 2011). Industrial works were divided into three classes based on their potential danger to the public with the class determining their proximity to towns and villages. Few processes were listed in 1810, but by 1867 they were extensive. Class 1 included public slaughter houses, manufacturers of fireworks and explosive, tanneries, and glue-works; Class 2 included soda works where the acid gas was condensed, blast furnaces, and potteries; Class 3 included breweries, brick works, paper manufacturers, and soap works. Inspectors were not appointed but enforcement was undertaken by the Préfecture de Department in each region of the country and by the Préfecture de Police in Paris. An important part of the procedure was the need for manufacturers to seek permission before the works was constructed. Unfortunately, enforcement was lax, perhaps confirming Robert Angus Smith's later concern that civic authorities in Britain might be compromised by vested interests. Later inspectors were appointed in the regions, while in Paris procedures were adopted for each district.

WATER SUPPLY AND SEWAGE TREATMENT

The supply of clean, fresh water became a major public health concern in Britain and proved to be a long-standing challenge. Authorities at both the national and local levels struggled to improve water quality and to ensure adequate supplies kept pace with rising populations. The population of Manchester and Liverpool rose dramatically through the period and increased the demand for water (Table 5.2). Waterways – rivers, canals, and streams – were used to dispose of human sewage while also acting as a "sink" for industrial waste products.

These contaminated waterways were frequently used as sources of fresh water. The diligent investigations by John Snow in London pinpointed the source of a major cholera epidemic in 1849 when he discovered that there was a high incidence of cholera cases close to water pumps where people were getting their water. During the 1854 epidemic Snow's work revealed that water supplied from outside London (and free of sewage) resulted in many fewer deaths. This confirmed that cholera was being carried by water contaminated with untreated sewage. Snow's conclusion later influenced how water quality was assessed. It was generally accepted that water quality was a measure of nitrogenous (or albuminoid) matter in the water, but in the 1860s and 1870s a major dispute arose over the best analytical method that involved Edward Frankland, professor of chemistry at the Royal College of Chemistry, with his combustion method and James Alfred Wanklyn, a lecturer at the London Institution, with his permanganate method. Other prominent chemists joined both sides in the dispute (Hamlin 1990: 185–90). Later, Frankland became internationally recognized as the leading expert on water quality.

Many industries were also culpable for poor water quality by using waterways to dispose of dangerous process effluents (Table 5.3). The major rivers serving the industrial centers became so polluted and detrimental to health that in 1865 Parliament established the Royal Commission on River Pollution Prevention to review the condition of several rivers and to make recommendations as to the best way of reducing pollution. There was an initial expectation that this pollution could be reduced in as effective a manner as the alkali inspectorate had tackled the acid gas pollution, but this did not happen. In their evidence to

TABLE 5.2 Population Increase for Manchester and Liverpool

	Population (thousands)	
	Manchester	Liverpool
1801	75	82
1851	303	376
1901	544	685

TABLE 5.3 Main Industries and Their Pollutants

Source/industry	Pollutants
Chemical	Hydrochloric acid
	Sulfur waste
	Ammoniacal liquors
	Nitric acid
	Sulfuric acid
	Organic residues
Bleaching	Bleaching powder
	Hydrochloric acid
	Sulfuric acid
Paper	Alkali liquors
	Rags
Tanneries	Spent tan liquor
	Lime liquor
Cotton	Bleaching powder
	Lime
Dye works	Organics residues
	Cyanides
Gas works	Tar residues
	Ammoniacal liquors
Agriculture	Fertilizers
Engineering	Acids
	Oils
Metal refining	Arsenic
	Cadmium
Woollen works	Vat liquors for printing
	Nitrogenous organic waste
	Soap residues

the Commission, industrialists sought to place the principal blame on human sewage, whereas the local authorities focused blame on the industrialists. While a method of treating industrial waste was often available (or could be sought), the necessary treatment plant required adequate space. Because most businesses operated in very cramped conditions, such extra space was not available, nor likely to be in the immediate future. The rate of improvement proved very slow and in an attempt to speed up improvement (while inspecting other rivers) a second Commission was established in 1868. After many delays, parliament finally approved the 1876 Rivers Pollution Act, but although there remained the challenge of treating both human sewage and industrial effluents, this legislation set a course of gradual improvement.

Deep wells were also often used to meet the demand for clean water across Britain, but by the 1850s these were proving inadequate. Manchester and Liverpool built reservoirs – Manchester at Longdendale (north Derbyshire) and Liverpool at Rivington Pike (West Pennine Moors). Further expansion in both locations was limited, so attention focused on the Lake District, an area of outstanding natural beauty some 100 miles to the north of both towns, where several large lakes offered a possible resolution. Manchester Corporation sought a parliamentary bill to use Lake Thirlmere and connect the town via a 96-mile aqueduct. Thirlmere's water was analyzed by the Manchester chemist Henry Roscoe, and found to be of very good quality. The bill laid out plans to build a dam at one end and raise the water level, and this provoked a public campaign against the proposal that included John Ruskin and Octavia Hill. The bill's subsequent passage through Parliament proved contentious, probably because each MP saw an opportunity for his own town and constituents. Nevertheless, the bill received the Royal Assent in May 1879 and Thirlmere began to supply Manchester with water in 1894. Later development of the project increased the capacity, so by 1915 the potential supply from Thirlmere was 30 million gallons a day against an average daily demand of 45 million gallons.

Liverpool had at one time hoped to share water from Thirlmere with Manchester, but this proved unworkable and instead Liverpool Corporation looked to the Vyrnwy valley in mid-Wales. There were initial concerns about water quality, but Edward Frankland confirmed its satisfactory quality. The Vyrnwy scheme also proved controversial because it involved constructing a large dam and flooding a valley. After Parliamentary approval in 1879, work began in 1881 and was completed in 1888, with water carried in a 68-mile aqueduct to Liverpool.

Urbanization and industry also posed similar challenges to cities and towns in the United States. Philadelphia became the first city to provide adequate water with construction of the Fairmount Water Works in 1802 (Tarr 1996: 9). New York, Boston, Detroit, and Cincinnati followed, so that by 1860 water works served the sixteen largest cities using 136 network systems. By 1880 the number of systems had increased to 598. The availability of water supplies increased both the number of users and the quantity of water; Chicago's consumption, for example, increased from 33 gallons per capita per day in 1856 to 144 in 1882.

Until 1855 no city in the United States had installed a sewer system, so waste water continued to run into local cesspools or into the street. The wide installation of water closets (but without a sewer system) exacerbated the already unhealthy conditions. This changed under the sanitary movement that had begun in Britain and was taken up by American cities from the 1850s leading to the installation of sewage systems in Brooklyn (1855), Chicago (1876), and Jersey City (1859; Tarr 1996: 237). Intense debates focusing on land use for treatment plant, costs, and the likely burden on taxes delayed the construction of many other systems, although construction accelerated after the 1870s. By 1908 cities with populations over 10,000 had 8,199 miles of sewers and by 1909 cities with populations over 30,000 had 24,927 miles (Tarr 1996: 12).

CHEMICAL HAZARDS IN THE WORKPLACE

Even in Britain with its advanced industrialized economy, workplace safety was a late development. The first thorough survey of chemical hazards in the workplace was conducted by Dr. Edward Ballard over the period 1876–1879 for the Local Government Board that had among its responsibilities health and the alkali inspectorate. Ballard looked at all industries including agriculture and provided a summary for each of both the nuisances and their origin. He also assessed their effect on the health of workers to provide a classification scheme for each industry or manufacturing process. Manufacturers were often ill-informed about the nuisances in their own factories or preferred to ignore them (Ballard 1878: 119). They did not keep a record of absences or illnesses. There was no monitoring of workers' health. The link between working with a particular process and the attendant health risks lacked scientific rigor. While lacking robustness, Ballard's report certainly drew attention to the scale of occupational health issues in Britain. Two examples from Ballard's study illustrate the serious health dangers for workers: the alkali industry (salt-cake workers), and the match industry (phosphorus necrosis).

The British Leblanc alkali industry employed about 12,000 people in 1890 when the United Alkali Company was formed by the amalgamation of 48 alkali works and salt works (Haber 1969: 183). Like most factories the alkali works were dirty and surrounded by waste products such as sulfur waste. As we have seen, the first stage of the Leblanc process produced large quantities of hydrogen chloride gas. The men operating this stage were called salt-cake men because the product was salt-cake (sodium sulfate) and their shifts were about eight hours. Salt-cake men were easily recognized by their teeth, which "if not entirely destroyed, are but black stumps. The effect makes itself seen in under twelve months" (Sherard 1896: 52). A bottle of whisky or brandy was always close-by in case of emergencies such as fainting or collapse. In such circumstances it is surprising that the Chemical and Copper Workers Union was not formed well before 1899.

The production of matches created problems of a different sort. From 1830 the white phosphorus used as a substitute for antimony sulfide proved to be extremely poisonous, and the workers producing these matches (most often women and young girls) suffered from phosphorus necrosis or phossy-jaw, which was debilitating and deforming (Crass 1941b). Although the safety match had been developed in the 1840s, the practice of using white phosphorus continued. In 1884 the "match girls" at the Bryant and May factory in London (one of the largest match-making factories in the world) went on strike to draw attention to their unhealthy working conditions. White phosphorus was banned in several countries between 1872 and 1901, and after the Berne Convention banned the use of white phosphorus in 1906, many more countries passed their own legislation, including Britain with the White Phosphorus Match Prohibition Act in 1908. The United States increased the tax on white phosphorus matches in 1913, making them uneconomical (Crass 1941b).

In Britain by the 1890s, mortalities were being collected for different occupations and for different age groups. These data reinforced the need for further parliamentary action, and in 1895 the Factory and Workshop Act required the notification of industrial diseases for the first time and designated additional inspectors, Chief Inspector of Factories (1896) and Medical Inspector of Factories (1898; Reed 2012: 149).

CONCLUSION

Chemistry and the emerging industrialized society were both radically different in 1914 than in 1815. Through the century more nations grew as industrialized societies and faced the consequences of urbanization and industrialization, both the positive benefits for consumers of commodities but also the damaging effects on the different elements of the environment. Chemistry had contributed to both, while also playing a major role in the advance in medical understanding that raised the longevity of life and benefited the general fight against disease. The events of the nineteenth century set in motion further protective regulations to control pollution and adulteration of food and beverages, and ensure safer air and water. The period also brought the large-scale availability of further chemical products for consumers and for international trade. Women had taken up wider access to education and shown they could play a broader and more fulfilling role in national affairs if given the opportunity. Toward the end of our period, however, there was increased concern on the part of governments and scientists over the future role of science (especially chemistry) in warfare and how the very nature of warfare and its scale of destruction might intensify. The events of World War I only confirmed these fears.

CHAPTER SIX

Trade and Industry: *New Demands, New Processes, and the Emergence of Science-Based Chemical Industry*

ANTHONY S. TRAVIS

INTRODUCTION

Over the course of the nineteenth century, society changed radically in the industrialized nations. Early in the century, the challenges posed during the Napoleonic wars by shortages of saltpeter, for gunpowder, and cane sugar in France testified to the vital role played by many products that were essential to both civil and military life. In the wake of these wars, the countries of Europe and the United States stood on the brink of a complete transformation of their economies through a new phase of the Industrial Revolution. By the mid-1820s, large quantities of a limited number of industrially produced chemicals were used in countless factories to produce a constant flow of material goods, ranging from textiles, to iron and steel, glass, bricks, and soap. The migration from rural areas and resulting growth in cities with the accompanying demand for housing created tremendous demand for the materials needed to build dwellings for the rising population of those seeking work in the new factories. The new urban workforce would need textiles for clothing, soap to combat increasingly unsanitary and crowded conditions, and increased agricultural productivity from the same amount of land, but with fewer farm workers.

The scale and speed of this upheaval were astonishing. For example, the population of Manchester, a city which typified Britain as the workshop of the world and whose fortunes arose out of the cotton industry, increased from over 70,000 in 1801 (with adjoining Salford) to over 300,000 in 1841 (Kargon 1977: 2). The chemical industry and its products played leading roles in all these emerging enterprises (Morris et al. 1988: vii). Among the greatest forces of change was the application of both practical and theoretical chemical knowledge in response to new needs and for the improvement of existing manufacturing processes.

Just after mid-century, the chemical industry had become central to the interplay of demand and change that enabled Europe to become the most economically dynamic continent (Haber 1969; Morris et al. 1991; Homburg et al. 1998). Around 1870, chemical industries assumed national importance in some countries. By 1900 the United States had caught up, and other nations, particularly Japan, were improving their chemical economies. In 1913, the chemical industry was dominated by four countries, representing together over 80 percent of world output of industrial chemicals: 36 percent from the United States, 25 percent from Germany, 12 percent from Britain, and 9 percent from France (Eichengreen 2003: 266–7). Germany, through its exceptional concentration on science-based industrial modernization, including self-sufficiency programs, had established the defining characteristics of twentieth-century chemical industry.

Coal provided the energy and the main raw materials to enable these transformations. Coal-fed furnaces drove iron machines and provided the heat that brought about useful chemical changes. Coal was heated in the absence of air to generate coal gas for lighting, or to produce coke, required in the production of iron and steel. From just after mid-century, the waste from these processes, known as coal tar, became the raw material for the production of organic (carbon-containing) chemicals, such as dyestuffs. In addition to the needs of the textile industry there was tremendous demand for agricultural fertilizers, and for metals, paints, and lubricants used in the railroad industry, which early on employed skilled chemists for analytical and other purposes (Russell and Hudson 2012).

A significant feature of growth in the chemical industry was the role of entrepreneurs whose technical skills counted as much, and sometimes more, than academic knowledge of chemistry. In many cases the entrepreneurs and inventors were young, and often backed by family members. In certain instances it was the combination of technical knowledge of inventors and capital from personally acquainted entrepreneurial families that enabled success. Despite hailing from different backgrounds, the entrepreneurs revealed surprising similarities in their approaches to problem-solving, and were united by networks of personal and professional relationships and their shared interests in manipulations of chemical materials.

No less important was awareness based on prior experience in or knowledge of a user industry. This was especially the case when the two main sectors of chemical industry emerged, one based on "fine" chemicals such as dyes, the other on "heavy" chemicals, such as mineral acids and alkalis. The growing textile industry provided the stimulation for growth in the production of both kinds of chemicals. By around 1870 rough and ready ideas and craft-based skills were giving way to the application of theory-driven chemistry, and as a result in many cases producing value-added products. A good place to begin an overview of all of these transformations is Great Britain, the cradle of manufacturing industries that relied on bulk supplies of acids and alkalis (Russell 2000b).

SULFURIC ACID AND LEBLANC ALKALI

Industries well outside of chemical industry invariably determined the specific need for most chemicals. Sulfuric acid and alkali (soda) are cases in point, and exemplify both the elaborate system of supply chains involving raw materials and intermediate products and the contingencies related to reliable supplies, especially of imported raw materials. By 1815, initially in Birmingham, sulfuric acid was manufactured in bulk from burned sulfur (sulfur dioxide) in lead-lined chambers, instead of expensive glass containers; the product served various trades engaged in metal working, or for pickling or cleaning. Because highways were poor and sulfuric acid was dangerous to transport, the corrosive acid was invariably manufactured near to where it was required, which is why, for example, it was produced in close proximity to the important tinplate industry of South Wales.

From the 1820s, the major consumer of sulfuric acid was the Leblanc soda-making (alkali) industry, based on a French invention, which first took root in northeast England. William Losh of Newcastle, who came from a family with interests in coal mines, had assisted the inventor Archibald Cochrane, ninth Earl of Dundonald, in developing the Leblanc process, based on readily available common salt and sulfuric acid (Clow and Clow 1952). They began trials in 1807 at Walker, Tyneside, near Newcastle. Dundonald soon departed and the business continued as the Walker Alkali Works, located next to, and associated with, a local iron works. Following extensive development, the successful manufacture of alkali was achieved by 1816. However, only after the abolition of the salt duty in 1823 did the Leblanc industry begin to flourish in Britain.

Close to the Losh works was the Newcastle Chemical Company of Christian Allhusen, at Gateshead, which opened in 1840, and was involved in alkali, and later glass, soap, and sulfuric acid production. It was at the time one of the world's largest chemical factories. The alkali industry also emerged in Scotland. In 1797, near Glasgow, three partners, Charles Tennant, Charles Macintosh, and James Knox, opened the St. Rollox works to manufacture aqueous bleach by passing chlorine through limewater. Two years later Macintosh's process for making a cheap bleaching powder was started. As a stable solid, bleaching powder (calcium hypochlorite) was easier to deliver to local textile factories than aqueous bleach. The powder enabled cloth bleaching in a few hours, replacing the bottleneck in textile processing that previously required exposure of cloth to buttermilk, sun, and air for long periods. Success led to the founding of the Leblanc alkali manufacturing firm Charles Tennant & Co. in 1814, and by 1835 it was the most important chemical works in the world. In 1842, a tall chimney over 400 feet high was erected at the works, in order to disperse the waste corrosive hydrogen chloride gas into the atmosphere, sparing the immediate environment but spreading the damage over the surrounding countryside (Morris 2003).

In northwest England, Lancashire became a major manufacturing center for alkali, driven by the rapid growth of the cotton industry, and again stimulating local manufacture of sulfuric acid (Campbell 1971: 29–31). In Liverpool, the Tennant firm opened an office offering alkali to soap producers and the glassmaking industry. Another leading entrepreneur in the Leblanc alkali industry of the northwest was James Muspratt, who in 1823 opened a factory in Liverpool. The Muspratt family erected a second Leblanc factory midway between Liverpool and Manchester, but complaints attendant on the release of toxic waste led to closure of both of their works in 1849. The Muspratts moved a few miles away to Widnes, on the River Mersey, close to St. Helens, a center of coal mining (Reed 2015). With its good canal and railroad connections to Liverpool and availability of salt from nearby Cheshire, Widnes became the main center of alkali manufacture and the site of the greatest concentration of British chemical industry (Figure 6.1).

The alkali manufacturer William Gossage successfully treated the waste hydrochloric acid gas produced in the Leblanc process by designing a tower in which the gas was dissolved in water. However, there was no demand for the resulting dilute aqueous acid, so it was dumped into nearby waterways. Nevertheless, the tower came into general use in the 1860s, by which time there were at least five Leblanc factories in Manchester and its environs.

The demand for alkali obviously led to a similar demand for the sulfuric acid required in its manufacture. In 1843, the German chemist Justus Liebig went as far as to write that "We may fairly judge of the commercial prosperity of a country from the amount of sulfuric acid it consumes" (Liebig in Clow

FIGURE 6.1 The cluster of alkali and allied chemical factories situated in Widnes, Cheshire, northwest England, close to the River Mersey and canals, which facilitated transport and in many cases acted as waste sinks. The several tall chimneys dispersed corrosive gases into the surrounding atmosphere. Probably late nineteenth century. Courtesy of the Sidney M. Edelstein Center, The Hebrew University of Jerusalem, Israel.

and Clow 1952: 130). The acid would become essential to the manufacture of so many chemical products that its output became a bellwether of a nation's industrial development.

In manufacturing sulfuric acid, the most important ingredient was sulfur, obtained mainly from Sicily. In the 1830s, John Tennant of the St. Rollox Works and his friend James Muspratt formed a partnership to develop Sicilian sulfur mines. However, in 1838 Ferdinand II of the Kingdom of the Two Sicilies, acting in collusion with two French businessmen, decided to profit from the business by enacting a royal decree that placed controls over sulfur production and exports. This, however, was an infringement of treaty rights with the British, and sparked a major diplomatic incident. In 1840, threatened with bombardment by a British naval fleet that had rushed to the Bay of Naples and closed it off, the king backed down (Kutney 2013: 51–9).

The uneasy dependence on foreign supplies of minerals encouraged the invention of processes using alternative sources. During the Anglo-Neapolitan crisis Thomas Farmer in London produced the intermediate, sulfur dioxide, from pyrites (iron sulfide). Britain's dependence on the Sicilian monopoly on sulfur ended when John Tennant and other leading alkali manufacturers purchased the Tharsis Pyrites Mines in southern Spain and in 1866 formed the Tharsis Sulphur & Copper Company.

Sulfuric acid manufacture was taken up on a large scale in the United States, where the supply of sulfur underwent a major change in 1894, when the German–American chemist Herman Frasch developed a new process for extracting native sulfur located deep below the surface in southern Louisiana. This involved sending superheated steam into the ground to melt the sulfur, then forcing it to the surface with compressed air, where it was solidified and then loaded onto railroad cars. This domestic production of sulfur led to a steep decline in the need for imported sulfur to the United States. It also enabled the United States to become the world's leading producer of sulfuric acid, and of equipment for its manufacture, by 1900 (Haynes 1959).

CONSUMERS OF ALKALI: SOAP AND GLASS INDUSTRIES

The production of alkali, first by the Leblanc process (and from the 1860s by the Solvay process, see later) transformed the manufacture and supply of soap. Two main types of soap became available: a hard laundry soap, and a milder household soap for personal washing and bathing. In the 1860s William Valentine Wright in London introduced the antiseptic Wright's Coal Tar Soap, which was used to treat skin disease (Campbell 1971: 68–74). Among the leading five or so British soap manufacturers was the Lever Brothers firm founded in 1885 at Warrington. They used a new process for a lather-free soap, called Sunlight Soap. The Lever workers were housed in a company village named Port Sunlight, and several subsidiaries were opened in other countries. The subsidiary in the Belgian Congo, a source of the palm oil that was essential for many new soaps, was notorious for its use of forced labor. Among other British leaders was the Pears soap business, where Thomas Barrett became a pioneer in consumer advertising through effective promotion of the healthful effect of Pears Transparent Soap. British-made soap was in high demand and exported to destinations throughout the world (Campbell 1971: 68–74). Other important firms in the soap business included the Gossage Soap Company, of Widnes, the Runcorn Soap & Alkali Company, and Joseph Crosfield's soap works in Warrington. During 1908–1909, the Cooperative Wholesale Society established factories in London and the northeast to manufacture soap in direct competition to the Lever Brothers (Wilson 1970).

Glass production grew mainly in areas such as Lancashire and the northeast, with its abundant supplies of coal and sand, and relied on alkali from Leblanc factories. The demand for window glass was enhanced by a significant reduction in the window tax in 1823, leading to an increase in the number of windows during the 1822–1825 building boom (Barker 1977: 28). In 1826, the St. Helens Plate Glass Company, at St. Helens, in Lancashire was formed, based on the technical knowledge of John William Bell, and a group of entrepreneurs consisting of three local families, including the Pilkingtons. By 1829, William Pilkington had taken control. In 1864, Pilkington set up Mersey Chemical Works to manufacture Leblanc alkali. Recovered chlorine enabled Mersey Chemical Works in around 1870 to produce 2,000 tons of bleaching powder each year.

Demand for alkali grew further with the need for bottles, and for large panes of glass for factory windows, railway stations, and other large buildings. Belgium was an important manufacturer of window glass, exporting over 30,000 metric tons of glass in 1860 and 213,000 tons in 1910 (Barker 1977: 442–3). A large American manufacturer of glass bottles was the Newark Star Glass Works opened in 1873, and succeeded in 1885 by the Edward H. Everett Company. By 1896 Everett was manufacturing 30–40 tons of glass bottles daily, with 500 men employed at the main works. The large Pittsburgh Plate Glass Company began operating in 1883.

SYNTHETIC DYES

In the 1850s, the British chemical industry was the largest in the world mainly due to demand from the textile industry for bleaching agents, dyes, and mordants, and from the glass and soap industries for alkali. Apart from dyes produced from natural substances, these industrial products were mainly inorganic chemicals, for which a number of improvements had been developed.

By the end of the 1850s a branch of the industry emerged based on carbon-containing compounds obtained from the waste tar available from coal-gas lighting works (Figure 6.2). These were the synthetic dyes, of which the first

FIGURE 6.2 Manufacture of coal gas by destructive distillation of coal. From mid-century the waste tar from the coal gas-making processes was transformed into synthetic dyestuffs, and the waste ammonia into ammonium sulfate fertilizer (Accum 1819). Courtesy of the Sidney M. Edelstein Library, The National Library of Israel.

was young William Henry Perkin's "aniline purple," first manufactured north of London in 1858 with family financial support. By the spring of 1859, this new color was all the rage among the fashionable ladies of London and Paris. Perkin then gave it a new name: mauve. In 1859, the Lyon colorist François-Emmanuel Verguin discovered a red aniline dye, fuchsine (magenta), and soon after in England a novel process for fuchsine was invented (Travis 1993). It was primarily fuchsine that sparked the rapid and remarkable transition in the textile industry from natural to synthetic dyes.

Trade in natural dyes involved exports of indigo from India to Europe (for dark blue), madder from the Near East (for bright red, and other colors according to the mordant), and dyewoods from South America, India, and Asia. By the 1830s, the growth of mechanized high-speed textile printing led to the need for improved extraction and application processes based on chemical techniques. The experts in the application and improvement of dyes were the textile colorists, who increasingly received formal instruction in chemistry while working as apprentices in the dyeing and printing industries. Academic chemists were drawn into investigations of the constitutions of natural dyes. Some chemists and colorists considered the possibility of creating artificial, or synthetic, colorants from readily available raw materials, notably coal tar, from which was distilled aromatic hydrocarbons such as benzene. Aniline, the intermediate required for production of mauve, and aniline red, could be produced from benzene in two steps.

Although Britain was the birthplace of the synthetic dye industry, its origins drew upon German scientific methods. When he made his invention of an aniline purple in 1856, Perkin was the eighteen-year-old assistant of the German chemist August Wilhelm Hofmann, who at the time taught at London's Royal College of Chemistry (Beer 1959; Garfield 2000).

THE WATERSHED OF THE 1860S AND 1870S

The changes that occurred between 1860 and 1880 were mainly the result of the introduction of synthetic dyes, and would characterize the beginning of the modern chemical industry. During this period, the dye industry was transformed from an artisanal enterprise, based on empirical methods in which practical chemical knowledge was applied, to one in which theoretical concepts and methods of chemical theory and synthesis were developed and applied. The industry that began in Britain and France attracted the interests of German entrepreneurs, in particular the textile colorist Heinrich Caro who in 1859, the year in which Perkin's mauve took off, moved from the Ruhr Valley to Manchester, where he joined the firm of Roberts, Dale & Co., which supplied chemicals to textile manufacturers. One partner, John Dale, a former student of John Dalton, recognized Caro's value to the firm. Caro's expertise in chemistry enabled him to invent novel dyes and apply them to dyeing and textile printing. Caro promoted the synthetic dyes among dyers and printers, mainly in Lancashire and Scotland (Reinhardt and Travis 2000: 93–4, 100–4).

Meantime, in London, by 1863 Perkin's teacher Hofmann was using theories of chemical constitution to design synthetic pathways. Although helpful in developing successful commercial dyes, these theories were not sufficient to fully explain the chemical structure of benzene and related aromatic compounds that were the basis of the new dye industry. The situation changed in 1865 when August Kekulé proposed that a six-membered ring of carbon atoms comprised the structure of benzene. This theory was put to use early in 1868 by two assistants of a former student of Kekulé's, Adolf Baeyer in Berlin. Carl Graebe and Carl Liebermann studied the important natural dye alizarin, the chemical basis of madder that was used in the long-established Turkey red dyeing process. They discovered, by stepwise degradation, that it was a derivative of the familiar aromatic compound anthracene. From anthracene they could then make alizarin, enabling them to infer a partial structure of the dye. Their novel method of synthesis was adopted by the new company BASF (Badische Anilin- & Soda-Fabrik), founded by a group of entrepreneurs, technical experts, and bankers. Other examples of such firms are AGFA (Aktiengesellschaft für Anilinfabrikation), Bayer, and the Hoechst dyeworks on the outskirts of Frankfurt. By 1865 BASF had located its factory complex for dyes and its in-house production of acids and alkalis in Ludwigshafen, just opposite Mannheim on the Rhine (Travis 1993). In 1868, Caro, who had returned to Germany, joined the company (Figure 6.3).

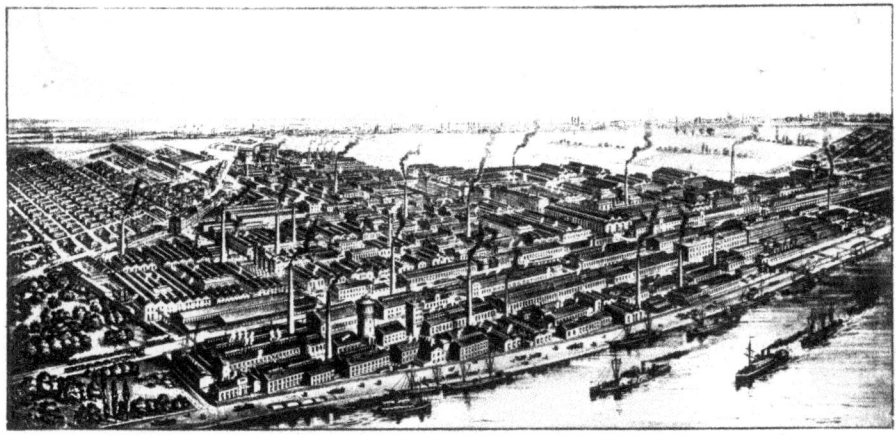

FIGURE 6.3 The Ludwigshafen factory of BASF (Badische Anilin- & Soda-Fabrik), located on the west bank of the River Rhine, late nineteenth century. In addition to synthetic dyes, the factory manufactured its own alkali and acids. Courtesy of the Sidney M. Edelstein Library, The National Library of Israel.

Caro is an outstanding example of how skilled individuals brought about transfer of technology from one district to another, or one country to another. At BASF, Caro established a commercial route to alizarin independently of and at the same time as William Perkin. The BASF and Perkin alizarin patents were filed in London just one day apart. Rather than engage in litigation over priority, BASF and Perkin agreed to share the market and exchange technical information. Industrial production began in 1869. The market for alizarin was controlled by a German–British cartel, the Alizarine Convention, one of the earliest major cartels in the nineteenth-century chemical industry.

However, Perkin soon fell behind in the technical development of the alizarin industry, and in the important elucidation of the total structure of alizarin. Baeyer and Caro achieved the latter in 1874, heralding the age of technoscience in which German chemists designed complex synthetic pathways and unraveled the constitutions of a number of colorants that had been obtained empirically. In 1875, Hofmann, in an attempt to overcome secrecy and promote protection of inventions by patents, published the basic structure of the new azo dyes, enabling an understanding of the method of synthesis, which was taken up by leading German dye firms. At the same time, dye chemist Otto N. Witt, then working in England, predicted, correctly, the color of an as yet unknown azo dye, enabling a theory of the connection between color and molecular constitution to be developed. As a result, both methods of synthesis and novel theories contributed to the invention of new dyes and their intermediates.

German chemists participated in the development of a uniform patent law introduced in 1877 that gave protection to science-based inventions throughout the German states. German firms vigorously protected their patent monopolies using an unrivalled mastery of theory and practice as demonstrated in the courtroom. Cases of patent litigation were not only fought against foreign firms that wished to evade German patents, but also between firms in Germany. One famous case over "Congo red" concerned a group of novel azo type colorants, the benzopurpurines, that were unusual and very valuable because they fixed directly to cotton without the aid of a mordant. In 1889, Caro as an ostensibly neutral consultant resolved a major dispute by drawing attention not to the product or process but to the new technical effect of the dye, thus expanding the scope of dye patents, which was accepted by the court. When AGFA won a case against Bayer over the benzopurpurines, Bayer did not appeal, but met with the opposition to draw up a cross-licensing arrangement and merge their interests through a cartel, which was allowed by the state in order to protect important industries (Reinhardt and Travis 2000).

The Germans were also aware of the need to maintain cordial relations with British manufacturers of important intermediates necessary for the production of dyes. Swiss dye-makers such as Ciba and Geigy, through imitation, benefited from the absence of a patent system that covered chemical inventions, at least until 1907 (Schiff 1971). Dye discovery involved the systematic synthesis of thousands of compounds, all of which belonged to a handful of categories. After the 1890s, few new classes appeared apart from a group known as vat dyes (first discovered at BASF in 1901), a range of greens and blues noted for brightness and fastness when attached to cotton.

Just as the synthetic dye industry was taking off, there was another major transformation closely connected to the textile industry's supply of chemicals. This was the appearance of the new ammonia–soda process for producing alkali, developed by Ernest Solvay in Belgium in the 1860s. This process enabled continuous rather than batch production, and was both far more efficient and far less polluting than the Leblanc process. In 1873, German-born Ludwig Mond, in partnership with John Tomlinson Brunner, erected a factory employing Solvay's process at Winnington Hall, near Northwich, Cheshire. This represented the foundation period of the prominent British alkali manufacturer Brunner, Mond & Company (Reader 1970; Morris 1989). In 1881, American licensees established the Solvay Process Company of Syracuse, New York (Bertrams et al. 2013).

American manufacturers of alkali expanded greatly in the early 1890s, and received a boost from the Dingley Tariff of 1897, which increased import duties on alkalis by 50 percent, and also introduced a new duty on imported bleaches. As was the case in Germany, such state-aided policies shifted chemical production away from Britain, which continued to rely on its free trade policy. However, in the United States there was one important exception, when in 1883 the specific duty on imported (mainly German) dyes was removed, in part to aid the textile industry. This did little to encourage the fledgling US dye industry.

In Britain, in order to confront the threat posed by the Solvay process, the United Alkali Company was formed in 1890 by amalgamation of 45 Leblanc firms, including Allhusen's, Mersey Chemical Works, and three salt producers. They managed to survive the onslaught of Solvay's process by making process improvements and saleable products from unhealthful waste. For example, Leblanc manufacturers employed the Deacon process to recover waste chlorine gas that was used for producing cheap bleaching powder, and beginning in the 1880s the Chance–Claus process was used for the recovery of sulfur to be used in the production of sulfuric acid (Warren 1988). In the long run, however, this was a losing battle. Following a long struggle for leadership, Solvay's process eventually replaced the highly polluting and more expensive Leblanc process.

MADE IN GERMANY

After 1870, Germany emerged as the leading nation in industrial chemistry; organic chemistry dominated chemical education and research, and there was a powerful and profitable relationship with industry. Marketing of dyes required qualified chemists who provided expertise on methods of application to industrial consumers. It also required special divisions, and hierarchies, particularly in management and marketing, and for the protection of patents. The monopoly on inventions provided to German firms by the patent system introduced in 1877 encouraged in-house research in dedicated research laboratories, often in individual manufacturing departments. All of these novel business practices, together with the central research laboratory, pioneered by Caro at BASF, became an important feature of science-based industries after the 1880s (Reinhardt and Travis 2000).

No less important were academic–industrial collaborations. In 1883, Baeyer (who worked closely with Caro) established the structural formula for indigo, and by 1890 a method of laboratory synthesis had been established. The natural blue dye had been extremely lucrative, used to color uniforms for the military, railway, and post office staff, and the clothing of most inhabitants of China. In 1897, both BASF and the Hoechst dyeworks began to manufacture the synthetic version, and the natural indigo industry started to decline, with significant economic consequences for the growers in India and their British investors. The British and French dye-making industries had also declined, mainly as a result of the neglect of science-based industry, despite appeals for revisions in patent law and improved science education. By 1900, the German firms, and to a lesser extent Swiss companies, dominated the global production of synthetic dyes, many of which, by suitable branding and colorful labeling of containers, were designed to appeal to indigenous and often illiterate peoples in South America, Africa, and Asia (Hoechst 1985).

PHARMACEUTICALS

The first major diversification of German dye-making firms involved conversion of coal-tar dye intermediates and their derivatives into novel pharmaceutical products. Until the early nineteenth century, curing diseases usually relied on a range of herbal extracts and inorganic mercury and antimony salts. The rise of anesthesia led to the introduction of the synthetic organic chemicals ether and chloroform. Preparatory work for both dyes and drugs was done in pharmacies. By the 1820s chemists had managed to isolate the effective agents from some of the natural products, in particular quinine from cinchona bark, which had been imported into Europe for treatment of malaria from South America since the seventeenth century.

The close connection between chemicals used to synthesize dyes and products of pharmaceutical value was established in the 1880s. In 1886, Kalle & Co., of Biebrich (Wiesbaden), made use of the accidental discovery by two medical researchers of the antipyretic effect of an aniline derivative, acetanilide, introduced under the trade name Antifebrin. At the Bayer dye firm, Carl

Duisberg attempted to convert a by-product from an aniline dye process into a derivative of acetanilide. The outcome was a product that showed analgesic and antipyretic properties, marketed as Phenacetin. This was Bayer's entry into drug discovery. In 1896, the firm opened a pharmaceutical research laboratory where in 1897 Arthur Eichengrün and Friedrich Hoffmann found that acetylsalicylic acid acted as an effective pain reliever. In 1899, it was introduced as Aspirin. In 1903 Bayer produced the first barbiturate sedative, Veronal (barbital). Heroin was another major, and highly profitable, product at that time. It was addictive, and became a widely used recreational drug (Sneader 2005).

Synthetic dyes also contributed to the study of how infections could be combatted within the body, notably by Paul Ehrlich (Travis 2008). In May 1909 with Sahaschiro Hata, Ehrlich discovered the curative action towards certain infections of a dye analogue named compound 606 (arsphenamine), which was marketed by the Hoechst dyeworks as Salvarsan. It was active in treating syphilis, and became called the "Magic Bullet," although there were side effects.

FEEDING GROWING POPULATIONS: FERTILIZERS

The improvement in the quality of soil by the addition of animal excrement, dung, urine, plant remains, fish, and general manures had been known for centuries, but during the nineteenth century chemical understanding of these fertilizers would change agriculture. Justus Liebig, a strong promoter of agricultural chemistry, was particularly impressed by the Chinese organization of the collection and use of manures in agriculture. Liebig recognized that the three most important elements required in plant growth were nitrogen, phosphorus, and potassium, and promoted the "law of the minimum," which stated that crop growth was limited by the least abundant of these nutrients. While Liebig's proclamations and claims were not always correct, they greatly encouraged the development of agricultural science. This appealed to British landowners and agriculturalists who were anxious to maximize crop yields to feed expanding populations. Landowners were the major supporters behind London's Royal College of Chemistry, opened in 1845, and headed by Liebig's star pupil Hofmann (Brock 1997).

To maintain the level of inorganic nutrients in plants, Liebig advocated using artificial sources of minerals. The modern fertilizer industry began with manufacture of phosphorus-containing "superphosphate of lime." In the 1840s, Sir James Murray developed a useful form of phosphate by treating crushed bones with sulfuric acid. Murray's business was taken over in 1846 by agricultural chemist John Bennett Lawes, who with Liebig's former student John Henry Gilbert had conducted field experiments with fertilizers on Lawes' estate at Rothamsted, Harpenden, England. This enabled Lawes to take the main initiative in commercialization, after filing a patent in 1846 for a superphosphate, and then starting its manufacture in London. By mid-century, phosphates were made from phosphate rock in Britain, Denmark, and Germany (Emsley 2000: 240–1). Later, manufacture was taken up elsewhere, including in Sweden, Italy, and Japan; superphosphate manufacture represented their first modern chemical industries. In the 1850s, the manufacture of superphosphate from coprolites (fossil dung) was taken up in England.

During the 1860s, Adolph Frank developed a new method for making potassium-rich potash. In 1857 Frank became a chemist at a beet sugar processing factory in northern Germany, where he learned about the newly emerging agricultural fertilizer business, in particular its role in improving the yield of sugar beets. Frank examined the waste salts available from the local salt industry, and developed a method for isolating potash from this waste that acted as an efficient fertilizer. Frank opened a potash factory in Stassfurt in October 1861. Soon after, he joined with and advised his rivals, who worked on improved extraction processes. Frank also promoted the German potash industry, and expanded the market to include the United States, which was an important exporter of potassium-containing wood ash (Travis 2018: 11–18). By 1870, the United States had become the largest market for Stassfurt potash, and the previously profitable American export of potash from wood ash quickly declined (Lodge 1938).

This new chemical approach to agriculture relied on agriculture research stations. In 1851, the first German agriculture experiment station was established near Leipzig. By 1890 there were over seventy such stations in the German empire (Harwood 2005). Similar stations were opened in Holland, Italy, Denmark, Sweden, and Norway. The first two US state agriculture experiment stations, founded on the German model, were opened in 1876, in Connecticut and California. The American Association of Official Agricultural Chemists was founded in 1884, under the patronage of the Department of Agriculture. Important activities included sugar analysis to detect adulteration and protect government and colonial coffers, which relied heavily on a system of sugar taxation (Warner 2011).

Liebig believed that nature supplied sufficient nitrogen by lightning strikes and excrement to enable crop growth. However, Lawes and others found that this was not the case, certainly not if rapidly expanding populations were to be fed. The need for bulk nitrogen fertilizer would be satisfied by nitrogen- and phosphate-rich guano, the accumulation of vast amounts of bird droppings found in the western part of South America, mainly the three Chincha Islands off the coast of Peru. Beginning in 1841, guano was first exported to England, from where most of the European market was mainly financed and regulated, and within a decade Europe was importing over 200,000 tons of

South American guano a year. From the late 1840s the British firm of Antony Gibbs & Sons monopolized the trade in guano through a contract with the government of Peru. In the 1850s, following investigations into the horrific working conditions forced on indentured Chinese labor engaged in extracting guano, the Gibbs firm took over the extraction side, although conditions only marginally improved (Leigh 2004; Melillo 2012). Guano created a fortune for the main partner, William Gibbs, who became the richest non-aristocratic man in England and built a vast estate at Tyntesfield in Somerset, close to the port of Bristol (Miller 2006; Kilburn 2009).

Such was the importance of guano that in August 1856 the US Congress passed the Guano Islands Act, empowering US citizens to take possession of unoccupied islands containing guano. In this early example of US imperialism, the president was empowered to send in armed military to intervene in cases of dispute, and it encouraged Americans to exploit deposits on islands in the Caribbean and Pacific. Over a hundred islands were claimed under the act, although most claims have now been withdrawn. The boom in the lucrative export business of guano created acute political tensions between Peru and neighboring countries, resulting in the Chincha Islands War (1864–1866), involving intervention by Spain, which failed to regain control of its former colony. The peak of guano production was reached in 1870, when 280,000 tons were imported into Britain (Travis 2018).

Guano was replaced as a source of fertilizer nitrogen by the mineral caliche, which contained sodium nitrate, from the dry, desolate Atacama Desert, then mainly in Bolivia, and to a lesser extent Peru (Figure 6.4). Nitrates were also essential for manufacturing the nitric acid necessary for the production of both synthetic dyes and modern explosives. Chilean investors exploited the nitrate deposits in the Atacama Desert of Bolivia using Chilean labor. In an attempt to gain income from the trade, Bolivia imposed punitive taxes on nitrate in February 1878, which created tensions with Chile. Bolivia was not willing to arbitrate, in accordance with a protocol of July 1875, nor was Chile willing to give in to Bolivian pressures. The outcome in April 1879 was the War of the Pacific, in which Peru sided with Bolivia against Chile. Chile was the victor, and claims were settled under the Treaty of Ancon, signed in October 1883, when Bolivia ceded its coastal territories to Chile, and Peru lost its nitrate deposits in the Atacama Desert and southern province of Tarapacá. As a result, Chile became the principal world supplier of nitrate for agriculture, explosives, dyes, and other chemical products. Until 1914, Chilean saltpeter, as the nitrate was now called, represented two-thirds of the world's supply of fixed nitrogen (Travis 2018).

From mid-century, sulfuric acid also played a new role in nitrogen fertilizer production. In this case the acid was used to convert ammonia-containing

FIGURE 6.4 Shoveling dry Chilean saltpeter (sodium nitrate), from evaporation pans, Arica, Chile. This photograph was taken in 1925, but the scene is typical of the large-scale nitrate industry in the late nineteenth century. Photograph by Bettmann/Getty Images.

refuse from coal gas works into ammonium sulfate; by 1890 140,000 tons of ammonium sulfate were produced each year out of the ammonia available from European coal gas and coke works (Hardie and Pratt 1966: 62). After Chilean saltpeter, this was the second main source of nitrogen fertilizer. Two-thirds of the ammonium sulfate made in Britain at gas and coke works was exported for use as fertilizer. From 1900, Japan was a major importer of ammonium sulfate, which was preferred to the Chilean nitrate for wetland agriculture because the nitrate was easily washed out of soil and formed toxic nitrite. Ammonium sulfate, eventually also obtained from other industrial sources, remained an important nitrogen fertilizer until the 1930s. By 1900, ammonia from gas works was used in industrial refrigeration, including in the meat-packing and brewing industries.

ELECTRICITY AND THE CHEMICAL INDUSTRY

Nitrogen-based fertilizer production was transformed during the first decade of the twentieth century by the direct fixation of atmospheric nitrogen (meaning converting it to its liquid or solid compounds) using electrical methods.

The processes required low-cost hydroelectricity, which determined new geographical sites of chemical production, mainly in Europe where the products were in greatest demand.

Using hydroelectricity to manufacture chemicals followed the opening of the first power stations in the 1880s. By the 1890s, electrical power was generated at Niagara Falls, and in Norway and Italy, and soon after in Japan. Chemical factories using this new major source of energy were located very close to the power stations, because long-distance transmission of electric current was then very difficult and wasteful. Early processes for electrochemicals such as aluminum were based on electrolysis. In 1886, Charles Martin Hall in the United States and, independently, Paul Louis-Toussaint Héroult in France reduced alumina to aluminum, eventually cutting the production cost 200-fold and making the lightweight metal inexpensive enough to use in making everyday items, such as tableware (Le Roux 2015). In 1897, the Castner–Kellner electrolytic process for caustic alkali was used in Cheshire, England, to supply chlorine for use in manufacture of bleaching powder. The production of sodium metal by electrolysis made possible cheap sodium cyanide, vital for the South African gold mining industry to extract gold from low-grade ores. This cyanide process, invented in 1887, was responsible for nearly a third of gold production in South Africa. By 1900, it was also widely used in Australia and the United States.

Electrolytic decomposition of low-grade ores transformed the production of copper, lead, and zinc, especially meeting the demand for copper in plating, locomotives, and electrical wiring for lighting systems. Electrolytic formation of copper was developed in 1869 by James Elkington at Pembrey, South Wales, and a similar process was developed independently in Hamburg, Germany, producing metallic copper in 99 percent purity. In the United States the output of copper reached 764 tons daily by 1903.

Other materials could be produced profitably from electrolysis. In 1884, for example, the Chemische Fabrik Griesheim in Germany began developing electrochemical processes, which in 1892 led to the production of potassium hydroxide and chlorine. In 1889, the Dow Chemical Company of Midland, Michigan developed an electrochemical process for obtaining bromine, and soon after adapted this to electrolysis of salt (sodium chloride) solutions to give caustic soda and chlorine (Whitehead 1968).

Electrochemical manufacture at Niagara Falls grew from the end of the 1890s, when for example Mathieson Alkali Works produced from its electrolytic chlorine a bleaching powder, prompted by the introduction of the Dingley Tariff. By 1910, Niagara Falls was the main center of the fast-growing US electrochemicals industry. In Britain the Leblanc manufacturers, combined as United Alkali, did not undertake the electrochemical manufacture of alkali. The Castner–Kellner electrolysis process for making chlorine replaced a clumsy and expensive method, while the chlorine byproduct enabled the development of chlorinated solvents and an early form of polyvinyl chloride (PVC); this further undermined the viability of the Leblanc process, because the electrolytic chlorine was much cheaper.

Regarding nitrogen, so crucial for both munitions and fertilizers, electrothermal processes for fixing this unreactive gas began after Sir William Crookes in 1898 predicted widespread starvation unless its fixation as stable compounds was achieved; he believed that electric arcs would solve the world nitrogen fertilizer problem. There were many technical difficulties, until 1903 when Norwegian physicist Kristian Birkeland, at the University of Kristiania (today Oslo), and engineer and entrepreneur Samuel Eyde fixed nitrogen as nitric oxide using an electric arc that replicated lightning. The oxide was readily converted into nitric acid from which the fertilizer calcium nitrate (lime nitrogen) was produced. For commercial development, Eyde attempted, unsuccessfully, to get financial backing from BASF, which was also interested in electric arc capture of nitrogen as part of its efforts towards diversification. On December 2, 1905, to work the Birkeland–Eyde process, Norsk Hydro was founded with mainly French capital. Shortly after, its first factory opened in Notodden. In 1911, the large Rjukan nitrogen facility was inaugurated. Electricity was generated at the Vemork power station, next to the Rjukanfossen waterfall on the Maan River. This, however, was at first a joint enterprise, involving BASF and Norsk Hydro (Sagatos 2005; Sogner 2014).

Another approach to nitrogen fixation that relied on abundant, cheap electrical power involved the reaction in electrically heated furnaces between calcium carbide and atmospheric nitrogen. This process arose out of the newly invented carbide processes. In 1891, the American Edward Goodrich Acheson discovered silicon carbide by heating a mixture of coke and quartz, and in 1894 sold it as Carborundum. In May 1892, Canadian inventor Thomas Leopold Willson heated lime with tar in an electric arc furnace and obtained calcium carbide, which with water produced the highly reactive gas acetylene. James Turner Moorhead commenced carbide manufacture in 1894, and four years later founded Union Carbide Company (Trescott 1989). Carbide-derived acetylene was an important source of lighting – with a flame far brighter than that from coal gas – used in lamps for homes, industry, city streets, cars, bicycles, and trains, and competed with electric lighting. Acetylene was also soon used for welding, the production of chlorinated solvents, and acetic acid.

The reactions of carbides involving electrothermal processes were investigated in Germany from 1895 by Adolph Frank and chemist Nikodem Caro. Dynamit AG (formerly Alfred Nobel & Company), of Hamburg, became interested in their work for producing cyanides. A factory was erected in Frankfurt, but was not a success, in part as a result of lack of demand during the Boer War. However, Frank and Caro found that the main nitrogen product from calcium carbide was a new product, calcium cyanamide, and Frank's son, Albert, demonstrated that cyanamide released ammonia on contact with soil. Thus, it appeared to offer potential as a nitrogen fertilizer. Calcium cyanamide production was first undertaken in Italy, from 1905, taking advantage of abundant hydroelectric power in the Abruzzo region (Travis 2018).

Large-scale manufacture in Germany began in 1908 at Trostberg, Upper Bavaria, and soon after factories were opened in Dalmatia, Norway, and elsewhere. Unlike the arc processes, the cyanamide process required pure nitrogen, leading to a close symbiosis between nitrogen production and the cyanamide process. The Norwegian cyanamide plant at Odda was claimed to be the first bulk user of nitrogen obtained according to the 1908 process of Carl von Linde for liquefaction and fractionation of atmospheric air. Nitrogen was also obtained by the Claude (French) process for liquefaction of air.

The Frank–Caro process was introduced into North America by Frank Sherman Washburn, who co-founded the American Cyanamid Company in July 1907 to work the process on the Canadian side of Niagara Falls. In Japan, calcium cyanamide production was undertaken from 1909 to 1910 by the forerunner of the zaibatsu, or industrial conglomerate, known as Nitchitsu. Cyanamide represented the most important early phase in the development of the modern Japanese chemical industry (Molony 1990).

The direct reaction of unreactive atmospheric nitrogen with hydrogen to form ammonia presented tremendous challenges, but would be met by the invention of the Haber–Bosch process. To study this problem, BASF hired a consultant, the physical chemist Fritz Haber, at Karlsruhe. In 1909, Haber, with the help of an English colleague, Robert Le Rossignol, demonstrated a small laboratory apparatus for continuous production of ammonia, under conditions of very high pressure and high temperature in the presence of an osmium catalyst (Stolzenberg 2004). Scaling up under Carl Bosch at BASF involved unprecedented engineering challenges. After thousands of experiments, Alwin Mittasch and colleagues eventually found an inexpensive catalyst, Swedish magnetite, which as it turned out was readily available directly from the laboratory shelf. When the new process for producing ammonia was described in New York during the summer of 1912, it met with great acclaim. Production began at Oppau, next to the Ludwigshafen works, in September 1913. The Haber–Bosch ammonia process, a tremendous technical achievement, was the pinnacle of a century of progress in chemical industry, although it became important only after late 1914, when applied in Germany to production of munitions (Smil 2001; Travis 2018).

NITRO COMPOUNDS AND EXPLOSIVES

Gunpowder, in addition to its military uses, was essential in civil engineering, particularly tunneling for railroads, and mining, and was also in demand by the sporting fraternity for shotgun ammunition. An essential ingredient for the black powder was potassium nitrate; manufacture relied mainly on imported nitrate from South America. Nitrates were also crucial for production of nitric acid used to make the explosive nitroglycerin, first manufactured by Swedish inventor Alfred Nobel. He opened factories near Hamburg in 1865, and, with investors from Glasgow, at Ardeer, Ayrshire, Scotland in 1871 (Fant 1993). The more stable dynamite, another nitroglycerin product, this time containing kieselguhr (diatomaceous earth), followed in 1875. Nobel developed a blasting cap that enabled explosions to be carried out under controlled conditions. His products enabled far-reaching changes in the scale and pace of mining and civil engineering projects. Alfred Nobel's other legacy is of course the prizes that commemorate his name.

Around 1870, the British explosives expert Frederick Abel developed a safe form of the blasting explosive guncotton, made from nitrocellulose. This was an important contribution to production of smokeless powders, such as the propellant cordite invented by Abel and Sir James Dewar and patented in 1889. Explosives that had slight shattering effects were employed not only in propellants but also for disturbing softer material, such as in railroad construction. The coal-tar derived aromatic nitro compounds picric acid (trinitrophenol) and TNT (trinitrotoluene) were adopted for military purposes after 1900. The manufacturing processes, similar to those in dye manufacture, required both nitric and sulfuric acids. In the United States, Du Pont, the leading manufacturer of dynamite and smokeless powder, became a monopoly through acquiring rivals, but in 1912, in accord with the Sherman Antitrust Act, was forced to divest its operations.

PRODUCTS FROM WOOD CELLULOSE

Commercial conversion of wood cellulose into pulp and paper was developed around 1840. The industry grew near large forests and rivers, notably in Scandinavia, the northwest of the United States, and in Russia. Production of paper relied on one of three chemical processes: the soda process, involving treatment with alkali, invented in the mid-1850s; the sulfite process, invented in Germany and Sweden, which involved heating wood with a mixture made up from the reaction of sulfur dioxide with calcium and magnesium compounds; and the sulfate, or kraft pulp process, a derivative of the soda process, which grew in importance after 1900. In addition, the finished paper products required supplies of soap, alum, dyes, and pigments.

Wood cellulose was also the starting material for a number of artificial products. The first was Alexander Parkes' Parkesine, a plastic-like material made by nitrating cellulose and treating the resulting nitrocellulose (a highly flammable material) with camphor. Following display at the 1862 London International Exhibition, it was anticipated that the product could substitute for ivory used in billiard balls, and in luxury goods, but commercialization was not successful, because of the explosive nature of nitrocellulose. In 1869 the American inventor John Wesley Hyatt filed a patent for making what became known as celluloid, also from nitrocellulose. Because of Parkes' prior work, Hyatt's patent was declared invalid. However, Hyatt's process was commercially a success, used, for example, as motion picture film and for replacing glass in photographic negatives (Friedel 1983). Cellulose also served as a starting material for artificial fibers, or rayon, based on processing of solutions of cellulose. First, in the 1880s, was with the nitrocellulose product Chardonnet, and then Bemberg silk. This was followed around 1890 by cuprammonium rayon. These processes relied mainly on the use of cotton linters. In 1904, the British Courtaulds company acquired the patents of consulting chemists Charles F. Cross and Edward J. Bevan to the viscose process for making an artificial silk from cheap wood pulp (rather than cotton linters). The cellulose acetate rayon was developed in America, from a viscous organic liquid, to give a synthetic fiber. The outcome was the American Viscose Corporation (General Artificial Silk Company), a subsidiary of Courtaulds, based near Philadelphia (and from 1911 at Marcus Hook; Blanc 2016).

FOOD ADDITIVES: SACCHARIN AND THE GROWTH OF MONSANTO

Coal tar chemistry contributed to a novel food additive discovered in the United States. In 1878, chemist Constantin Fahlberg, assistant of Ira Remsen at the Johns Hopkins University, Baltimore, found by chance that a newly synthesized compound was intensely sweet. It was named saccharin, manufactured in Germany by Fahlberg and a relative from 1887, and marketed as a substitute for natural sugar. Other German companies soon began producing saccharin, and in late 1901, the American entrepreneur John F. Queeny, a marketing man without a chemical background, founded the Monsanto Company to manufacture the sweetener in the United States. Unable to find American chemists sufficiently familiar with coal tar chemistry, Queeny brought in chemists from Switzerland. German manufacturers responded to Queeny's success by selling saccharin at an unsustainably low price "on the American market," with the aim of driving Monsanto out of business. In addition, the German companies even denied Queeny the essential intermediate. Queeny's chemists responded by devising a new synthesis of the intermediate, which removed their reliance on Germany.

The widespread use of saccharin raised concerns about its safety as a food additive. The US Department of Agriculture's Chief Chemist, Harvey Washington Wiley, lobbied for control of saccharin and other food additives. In 1907 he brought evidence before President Theodore Roosevelt, who was a regular consumer of saccharin. Roosevelt installed a referee board to investigate the properties and effects of the sweetener. Although strict regulations on use were introduced, the outcome was that Queeny was enabled to continue production. After 1914, Monsanto's process for the intermediate, now no longer available from Germany because of the war, enabled the firm to monopolize the US artificial sweetener industry (Warner 2011: 181–94; Forster et al. 2020).

MODERNIZING BUSINESS PRACTICES

The chemical industry in Germany, supported by patent monopolies on chemical inventions, was encouraged to expand research and development programs, and to construct in-house laboratories. No less important were the growing management teams, marketing divisions, and departments that supported litigation in patent disputes. As the chemical industry became increasingly globalized, there were moves toward collusion among manufacturers in different countries to control markets as well as to share technologies. Larger firms were often organized into syndicates, trusts, or cartels, sometimes to satisfy the demands of investors. Companies entered into joint ventures to fight off foreign or local competition. External threats, whether foreign corporations, rival processes, or warfare, stimulated consolidations, and monopolies, of influential producers, as well as alternatives to monopoly, particularly on natural nitrates (Blakemore 1974). In several cases, mergers were the outcome. In 1886 the Nobel–Dynamite Trust Co. was established in London to bring about a merger between Nobel's company in Germany and two British companies (Fant 1993).

In Germany, the leading dye firms in 1904 consolidated their interests by establishing two loose alliances: BASF, AGFA, and Bayer, together known as the Dreibund, or Triple Confederation; and Leopold Cassella & Co. and Hoechst, the Double Alliance, in 1907 joined by Kalle & Co. and from then on known as the Tripartite Association, or Dreiverband (Haber 1969). The first important merger in the United States took place in 1899 when twelve firms formed the General Chemical Company.

Associations of manufacturers emerged to undertake promotion and quality control and regulate prices. In 1872 the Manufacturing Chemists' Association was formed in the United States by a group of sulfuric acid manufacturers. The British Ammonium Sulphate Federation was founded in the 1890s to coordinate the substantial export market of the sulfate. In some countries, notably France, the state controlled the sulfate market that was critical to the development of agriculture (Travis 2018). The control of markets was invariably dictated by different political or economic interests in different countries, favoring home industries or encouraging exports, as the prevailing situations demanded.

CONCLUSION

An outstanding feature of the chemical industry in this period was the capacity of individual entrepreneurs to seize opportunities offered by developing technologies. Some, such as William Perkin and Ernest Solvay, were young, hardly out of their teens, when they developed their inventions as small enterprises in a yard or back shop. Their activities, often with partners who provided capital or marketing expertise, represented the pioneering epoch. In many instances the entrepreneurs relied on waste products, available at little or no cost, and on which some prior laboratory work had been done. They were both versatile and agile in their business practices, which sometimes included industrial espionage. Thus in 1845, the Danish entrepreneur Jacob Christian Jacobson surreptitiously carried home Bavarian yeast in a tin hidden in his hat box, and two years later established the Carlsberg Brewery near Copenhagen (Holter and Møller 1976).

Prior to the 1860s, Great Britain was the main source of capital for the new chemical industry. Apart from massive investments in chemical-based activities related to textiles, British investors supported and in some cases controlled the exploitation of foreign sulfur mines, the nitrogen and copper industries of South America, and the mining industry of Mexico. This comparative advantage, with supply chains globally integrated, was enabled by the empire's pre-existing maritime trading routes, and, in the case of mining, to the application of novel chemical processes in the processing of low-grade ores. Historical and geographical contingency also played an important role, particularly reliance on South America for fertilizers, on India for indigo dye (until the artificial dye went into production), and fast-flowing rivers for the development of hydroelectric power for producing new materials or making old materials more efficiently. Traditional industries that employed thousands of people were threatened by the obsolescence of old materials and techniques, such as natural dyes. As new sources of chemicals arose from science-based inventions, industries tended to move nearer to their main markets, unless sources of low-cost energy dictated otherwise, particularly hydroelectricity in Scandinavia, Japan, and the Niagara Falls region in North America.

By 1910, through products of chemical industry, the lives of millions were enriched with improved health and enhanced crop yields, and an abundance of material goods. Consumption was satisfied by novel methods of production: Synthetic products became substitutes for chemically identical natural products (madder and indigo dyes), completely novel substances without analogues in nature were available (dyes, saccharin, and artificial silk), and minerals and unreactive nitrogen were harnessed to the needs of agriculture. Industrialized nations saw the diversified chemical industry as a critical part of industrial policy. In the nineteenth century, the growth of the chemical industry altered regions and nations; as the chemical industry entered the twentieth century, its impact would become increasingly global.

CHAPTER SEVEN

Learning and Institutions: *Emergence of Laboratory-Based Learning, Research Schools, and Professionalization*

PETER REED

INTRODUCTION

During the nineteenth century chemistry was the first modern and the most readily applied branch of science that took a leading role in the broad integration of science in society. As regards education, universities showed remarkable growth in the emerging industrialized nations, transforming chemical pedagogy and creating institutions to foster professionalization of chemists. Outstanding chemists defined and directed a new laboratory-based method of learning. A small and distinguished group of these chemists became leaders of innovative research schools that focused on investigating specialized topics in chemistry. This research not only advanced chemical knowledge and understanding, but also brought benefits to industry, the economy, and their citizens' well-being.

The following discussion of the development of chemical pedagogy and institutions during the nineteenth century embraces six interrelated themes: the move away from the classical curriculum towards science; the emergence of the university as a center for research; the introduction of laboratory-based teaching; the professionalization of chemists; the steady fragmentation of chemistry into specialties with their own organizations and journals; and national differences influencing these developments.

LOCAL CULTURAL INSTITUTIONS

In Britain, the literary and philosophical societies (or "Lit and Phils," as they were often called) that had been founded in many towns during the second half of the eighteenth century established natural science as a principal focus through the nineteenth century. These institutions continued to provide a regular meeting place for local industrialists and merchants to pursue their literary and scientific interests through lectures and a well-stocked library. Members kept abreast of the latest news through newspapers and periodicals, and networked to develop new opportunities for business, science, and the arts.

A related development of the early nineteenth century was the mechanics' institute movement, which grew out of a concern for the education and intellectual "improvement" of workers in many industrialized countries. In Britain, institutes were often established by leading members of Lit and Phils, who recognized the benefits of educating their workers. The adult education programs (often in the evenings) had a focus on chemistry, in recognition of its increasingly important role across many sectors of industry. The first mechanics' institute in Britain was established in 1821 in Edinburgh, with Liverpool (1823) and London (1823) following soon after. They also emerged in other countries, including in Hobart (Tasmania) in 1827, Atwater (Quebec) in 1828, Sydney (NSW) in 1833, Newcastle (NSW) in 1835, San Francisco in 1854, and New York in 1858.[1] By mid-century there were estimated to be over 700 institutes across Britain and overseas. The Public Libraries Act of 1850 brought about major changes in Britain, with the transfer of many institute libraries to the new subscription-free public libraries. Further changes occurred through the second half of the nineteenth century, as newly established colleges and universities began offering evening classes as well as degree courses.

In London, the Royal Institution of Great Britain (RI) became an important venue for the popularization of science, and a major center for scientific research. Founded in 1799 on principles drawn up by Benjamin Thompson (Count Rumford), the American natural philosopher and philanthropist, the RI was established by private subscription in the center of London. In 1801, at Rumford's invitation, Humphry Davy was appointed assistant lecturer in Chemistry, director of the chemical laboratory, and assistant editor of the journals (James 2000: 2–5). He organized a series of spectacular and entertaining public demonstrations that both men and women attended; his demonstration of the physiological effects of nitrous oxide gas (laughing gas) drew large and excited audiences. In 1802 Davy was appointed professor of chemistry and by 1812 had established the RI as a center for research. Davy's own research in electrochemistry led to the isolation of several chemical elements, and in the period 1815–1816 he invented the miners' safety lamp that saved many lives.

In 1813 a young bookbinder's apprentice named Michael Faraday was appointed chemical assistant at the RI, having attended Davy's lectures and become interested in chemistry through reading Jane Marcet's *Conversations in Chemistry* (Figure 7.1). Faraday proved to be an outstanding experimental scientist and spent the rest of his working life at the RI, including initiating the Friday evening discourses and the Christmas lectures for young people – his series of lectures, *The Chemical History of a Candle*, being particularly celebrated (Faraday 2011 [1861]). Although most famous as a founder of electromagnetic science, his outstanding contributions to chemistry included the laws of electrolysis, the isolation of benzene, the liquefaction of gases, and improvements to steel and

FIGURE 7.1 Faraday's laboratory at the Royal Institution. Photograph by Picture Post/Hulton Archive/Getty Images.

optical glass. Faraday's standing as a scientist led to his role advising a number of organizations including the Admiralty and Trinity House (the organization responsible for lighthouses), and to his appearance as an expert witness in several major court cases. Both roles aided the growing professionalization of science (James 2000: 12).

It was only in 1862 that the RI formally and officially adopted research as a principal activity. John Tyndall, professor of natural philosophy at the RI since 1853, made many important research contributions, as did Edward Frankland, who was professor of chemistry between 1863 and 1868. To enhance the RI's research facilities, in 1896 the German-born industrial chemist Ludwig Mond provided an endowment to purchase the building at 20 Albemarle Street and create the Davy–Faraday Research Laboratory. The appointment of John William Strutt (Third Lord Rayleigh) and James Dewar in 1896 as joint directors of the Laboratory further strengthened this commitment to research. Rayleigh, together with William Ramsay at University College London, discovered the "noble" gas argon.[2] Dewar's work led to the invention of the Dewar vacuum flask.

A number of institutions in London and its environs adopted the model of the Royal Institution and were committed to scientific education and, to a lesser extent, research. These included the London Institution (1806–1912), Surrey Institution (1807–1823), and Russell Institution (1808–1881). The London Institution was the longest surviving and the most active in scientific education. Chemistry took a very high profile with the appointment of James Wanklyn and Henry Armstrong as professors of chemistry and with access to a small research laboratory. Other leading chemists such as Michael Faraday and William Ramsay also gave lectures there.

SCIENCE AND CHEMISTRY IN SECONDARY SCHOOLS

Concerns expressed after the Paris International Exhibition in 1867 over Britain's perceived industrial decline prompted a series of government commissions and select committees on science and technical education through the remainder of the century (Cardwell 1972: 111–19). Under review was the provision of education from elementary schools to universities; a special focus was on science and technical education, and the need to move away from the traditional educational philosophy that emphasized the classical languages. While raising concern about the general inadequacy of elementary and secondary education, the 1868 Report of the Select Committee on Scientific Education (or Samuelson Committee) drew attention to a lack of well-qualified science teachers and good laboratory facilities in schools, and how these deficiencies were responsible for the poorly trained students who entered universities.[3] The Samuelson Report was probably responsible for galvanizing support among MPs for the important 1870 Education Act that initiated a universal educational system with compulsory education to age ten.

In 1870 several prominent scientists were appointed to the Royal Commission on Scientific Instruction and the Advancement of Science (the Devonshire Commission). Its subsequent eight reports and four volumes of evidence took a wide sweep across many different aspects of education and science (Cardwell 1972: 119–26). The Commission recommended a radical reform of secondary education, and saw an urgent need for better science education. However, as with the recommendations from other reports, a laissez-faire approach to education remained the policy of the country. Successive governments proved reluctant to give education a high priority, and improvement only came in small steps over the rest of the century and into the next. Compulsory education was raised to age twelve by the 1902 Education Act, and to age fourteen in 1918 (with a recommendation for part-time education between fourteen and eighteen). As late as 1916 the Thomson Commission was drawing attention to the difficulties experienced by science teachers in many schools due to the hostility of headmasters and colleagues toward science (Roderick and Stephens 1972). Under the influence of Henry Armstrong, science education steadily adopted the heuristic method, in which children learned through experience and experiment (Brock 1993: 409).

In France, educational curricula and examinations had been set by Napoleon before 1814. The *lycées* (secondary schools) were established in 1802 under the influence of the scientist Antoine-François de Fourcroy, who stressed the importance of a detailed timetable for the pupils, the strict authority of the teacher, a disciplined approach, and an emphasis on rote learning. Such a format featured for the next hundred years. In 1808 a curriculum for secondary schools was defined that included the physical sciences, and the post-*lycée* examination, the baccalauréat, was established for the elite group who would go on to study in universities or gain positions in the civil service (Nye 1986: 12).

Prussia established its official matriculation qualification, the *Abitur*, in 1814 and by 1834 it was made the entry qualification for universities. Later in 1871, the *Abitur* was decreed for all German states and for entry to all German universities. The school system also became more centralized (as in France), and different types of secondary schools were established – the classical *Gymnasium*, the *Realgymnasium*, the *Realschule*, and the *Oberrealschule* – with different periods of education and different emphases but including science. With the growing demands of the professions and industry, the *Gymnasium* schools with their focus on the *Abitur* qualification became especially important.

In the United States (like Britain), an assault on the prevailing classical curriculum in schools during the nineteenth century led to the widespread introduction of science into the curriculum (DeBoer 1991: 17). By the 1890s, however, science was overloading a school curriculum that was also facing challenges arising from the different entrance requirements of colleges and universities. In 1890 only 360,000 fourteen- to seventeen-year-olds attended high school (about 6.7 percent of the age group) and yet by 1920 the number would rise to 2.3 million (32.3 percent; DeBoer 1991: 39). In July 1892 at the annual meeting of the National Education Association, a committee of ten college and school heads (known as the Committee of Ten) was appointed under the chairmanship of Charles Eliot, the president of Harvard University. A leading recommendation was that science should occupy up to 25 percent of the curriculum (DeBoer 1991: 41). Separate conferences were also convened to review specific curriculum areas; these included the Conference on Physics, Chemistry, and Astronomy under the chairmanship of the chemist Ira Remsen of Johns Hopkins University. Recommendations from the conference included requiring physics and chemistry for college admission, beginning instruction in physical sciences in the elementary schools, and basing half of the work in the physical sciences on laboratory work. The rapid rise in college science enrollments between 1890 and 1910 was attributed to work of the Committee of Ten (DeBoer 1991: 50).

CHEMICAL INSTRUCTION IN UNIVERSITIES AND COLLEGES

Between 1815 and 1915 most industrialized countries experienced a remarkable growth in the number of universities and colleges. In Britain and the United States most initiatives were at the local, regional, or individual state level. In Germany and France there was considerable state support for universities (or their equivalent), but while German universities were largely autonomous institutions of independent German states, universities throughout France were all part of a government institution, the *Université de France*, which was run by a bureaucracy in Paris and subject to political control. It was against this backdrop of opportunities and constraints that chemistry emerged as an important academic discipline in higher education, steadily embracing laboratory-based learning and the creation of research schools.

In Britain there were additional universities and colleges and greater access to higher education for the many students excluded by religious strictures. In 1815 Britain had six universities, two in England and four in Scotland. Higher education at this time emphasized the classics (Oxford), mathematics (Cambridge), and philosophy and medicine (Scotland). In Scotland, chemistry was taught as an integral part of medicine but not as a free-standing subject. The purpose of studying at these universities was the training necessary to become clergy, lawyers, teachers, doctors, and civil servants, but this situation was to gradually change, especially from the 1850s, with training for careers in the sciences or business (Sanderson 1975. 1). Many of the changes were local initiatives (often in London and the larger industrial towns), while others were implemented following parliamentary select committees and royal commissions. Science was to have a crucial role in this new higher education system, and chemistry would carve out an influential niche, although not without financial demands for well-equipped laboratories and experienced staff.

The 1820s saw the foundation of several new universities in Britain including University College London (UCL; 1826) and King's College (KCL; 1828), both in London, and Anderson's University (1828) in Glasgow. UCL was the earliest university to be founded in London. It was created privately in 1826 as London University, but when in 1836 the separate University of London was created as a degree-awarding body, the College changed its name to University College London. From its foundation UCL embraced chemistry, and the first chair in that science, Edward Turner, had access to a laboratory for teaching practical courses in chemistry. Such teaching laboratories were to become an important feature of higher education. The Scottish-trained chemist Thomas Graham followed Turner in 1837. Graham had been Professor of Chemistry at Anderson's University in Glasgow and was a strong supporter of learning chemistry through practical laboratory work. Both Graham and Turner published influential textbooks; Turner's *Elements of Chemistry* (1827) was widely used across Europe, and Graham's *Elements of Chemistry* was published in six parts between 1837 and 1841. Attracting prominent chemists such as Graham and later Alexander Williamson (in 1849), and construction of the Birkbeck Chemical Laboratory (the first purpose-built laboratory building in Britain) in 1845, helped UCL to become a leading university for chemistry in Britain.

In the 1840s there was concern over London's lack of instruction in organic chemistry, an important emerging field linked to pharmacy, agriculture, and other practical arts. With prime minister Peel taking a leading role, having sought the views of the leading organic chemist, Justus Liebig at the University of Giessen, a school for higher education in chemistry was founded in London in 1845 (Bentley 1970: 161). It drew on the strong links between British chemists and Liebig's growing network. Liebig's method of laboratory instruction had attracted many chemistry students from Britain (and many other countries) to come to Giessen and learn from the master (Fruton 1988: 32–8). For director of London's new Royal College of Chemistry, Liebig recommended his assistant, August Wilhelm Hofmann, who brought current research in organic chemistry and Liebig's teaching research model to Britain. Hofmann divided the laboratory space into discrete areas for students at different levels, but he also retained a personal role encouraging students and maintaining discipline (Jackson 2011: 57). The 1856 discovery of the first synthetic dyestuff, aniline purple, soon thereafter christened "mauve," by one of Hofmann's students,

William Henry Perkin, was justification for Hofmann's appointment and his laboratory teaching style.

In 1851 the Royal School of Mines was established in South Kensington, London, to encourage the teaching of science in Britain, following the success of the Great Exhibition (Bentley 1970: 156). The Royal School of Mines had evolved from the Museum of Economic Geology (founded in 1837). When in 1853 the government established the Department of Science and Art which would incorporate the School of Mines, Giessen-trained chemist Lyon Playfair resigned his chemistry post at the School of Mines to become Secretary for Science in the new Department, and Hofmann took over the vacant post, while the Royal College of Chemistry was incorporated into the School of Mines. In 1872–1873 the College of Chemistry moved into new premises in South Kensington, and in 1881 was redesignated the Normal School of Science to reflect its wider science education role (it was renamed the Royal College of Science in 1881). A further merger took place in 1907 when the Royal College of Science, the School of Mines, and the City and Guilds Central Technical College (Finsbury Technical Institute from 1893) formed Imperial College of Science and Technology that became a "constituent college" of the University of London (Argles 1964: 80–1; Gay 2017: 35).

While these institutions in London were evolving, many major regional towns aspired to have universities, and often drew on German influence (Haines 1958). Discussions about a university for Manchester, a major industrial and commercial center in the north of England, had taken place over many years, but it was not until 1846 when John Owens, a wealthy Manchester merchant, included a donation of £96,954 in his will that the foundation of a college or university seriously moved forward. Owens College opened in March 1851 and became a model for the civic university movement in Britain (Kargon 1977: 157). The first chair of chemistry (a part-time appointment) was Edward Frankland, who had studied with Robert Bunsen at Marburg and Liebig at Giessen. A chemical laboratory was added after additional funds were secured. Following Frankland's resignation in 1856 (possibly due to the low student numbers in the early years), Henry Roscoe, who had studied with Graham and Williamson at UCL and with Bunsen at Heidelberg, was appointed to the chemistry chair, a post he retained until 1885 when he was elected Member of Parliament for South Manchester.

Roscoe undertook consultancy work for local industrialists and merchants to demonstrate the value of studying chemistry at Owens College, and working alongside other staff advocates, the prospects for Owens College steadily improved with the rise in student numbers (Jones 1988: 51). In 1873 Owens College moved to new buildings on Oxford Road designed by Alfred Waterhouse. Roscoe designed new chemistry laboratories and went on to create an outstanding school of chemistry based on his experiences studying at

THE ROSCOE LABORATORIES.—QUANTITATIVE.

FIGURE 7.2 The qualitative analysis laboratory, Owens College, Manchester, 1873 (Roscoe 1906). Courtesy of Peter J.T. Morris.

Heidelberg (Figure 7.2). Roscoe attracted outstanding students who would go on later to senior appointments in academia and industry, including William Dittmar (later at Anderson's University, Glasgow) and Thomas Edward Thorpe (later at the Royal College of Science and the Laboratory of the Government Chemist). In 1874 Carl Schorlemmer was appointed to the first chair in organic chemistry in Britain. Besides its outstanding research output, the faculty in the department published a wide range of textbooks, including Roscoe's *Lessons on Elementary Chemistry* (1866), *Primer of Chemistry* (1871), and *A Treatise on Chemistry* (from 1877 in several editions) with Schorlemmer, who also wrote *A Manual of the Chemistry of Carbon Compounds* (1874); Francis Jones published *Owens College Course of Practical Chemistry* (1872).

However, Owens College aspired to become a university that could award its own degrees, having had to rely on examinations and degrees administered by the University of London. In 1880 the charter was granted for the Victoria University, with Owens College as the only constituent college. Later, other colleges joined the federal Victoria University, namely University College Liverpool in 1884 and Yorkshire College Leeds in 1887, but this arrangement proved temporary. Each College sought independent university status with the ability to award its own degrees, and this change in status was achieved over a period of just six years: Victoria University of Manchester (1903), University of Liverpool (1903), and the University of Leeds (1904). The University of Birmingham (1900) and University of Bristol (1909) had also achieved university status.

The 1871 University Test Act finally removed all restrictions on degrees and awards at Oxford and Cambridge universities for dissenters and non-conformists (Sanderson 1975: 32). Chemistry as an academic field was to develop in different ways in the two universities (Sanderson 1975: 148–51). At Oxford, the major steps were the establishment of the Honours School of Natural Science in 1850 and the building of the University Museum in 1855 to house scientific collections and to provide lecture rooms and laboratories for teaching science, although individual colleges had their own laboratories. An additional building for inorganic and organic chemistry was erected in 1876. In 1865 Benjamin Brodie was appointed the first Waynflete Professor of Chemistry; he was succeeded by William Odling in 1872. William Henry Perkin Jr. was appointed to the chair in 1912 and was also the first head of the Dyson Perrins Laboratory, completed in 1916 with financial support from Dyson Perrins of the Worcestershire Sauce family. The Laboratory was devoted to organic chemistry.

At Cambridge students were only able to get a degree in chemistry following adoption of the Tripos examination scheme in natural sciences in the 1860s. This allowed a student to study a broad base of science in Part 1 and then choose a more specialized field (e.g. chemistry) in Part 2. Although laboratories in some of the colleges existed, the first University Laboratory was constructed in 1887 (Freemantle 2002: 39). Cambridge had a Professor of Chemistry from 1702 and there were three appointees over the period 1815–1915: James Cumming who was appointed in 1815, George Liveing in 1861 who worked on spectroscopy, and William Pope in 1908 whose work focused on stereochemistry (Archer and Haley 2005). Pope was credited with advancing the role of research in the department (Freemantle 2002: 40).

In the German states, many universities had existed prior to 1810, but the foundation in Prussia of the Universities of Berlin in 1810 and Bonn in 1818 exemplified reforms by Wilhelm von Humboldt (brother of Alexander von Humboldt) that embraced Enlightenment as well as neohumanist principles, and put the unity of research and teaching as its higher priority. Research was thereafter included in the reforms of other universities across the German states. A huge wave of building university laboratories and institutes followed in the decades after the revolution of 1848. The perceived German scientific and technical strength at the International Exhibition in Paris in 1867 prompted France, Britain, and the United States to review their own higher education policies and to make research a major objective. However, in addition to the humanistically influenced secondary and higher education, the German system also had its *Technische Hochschulen*, which focused on engineering and applied science, originating as a response to the mass industrialization associated with the Industrial Revolution. These institutions were widely spread, like the universities, across German-speaking Europe.

Qualifications played a major role in gaining access to the different levels of education in Germany, and there was a very strict route for entry to universities from secondary education. Passing the *Abitur* examination at the *Gymnasium* was the prerequisite for studying at universities, for taking other higher-level examinations, or to gain entry to the mid-level of the civil service. Universities were able to confer two degrees: a doctoral degree for research based on a dissertation, and the *Habilitation*, a second degree that granted the right to offer lectures at a university, but without an official salary. Despite their dependence on government funding, universities were largely autonomous and authorized to award degrees. The emphasis on research separated universities and the *Technische Hochschulen* until 1899, when the latter were permitted to award doctoral degrees.

In the last half of the nineteenth century, individual German states used the inducement of new state-of-the-art laboratories to attract outstanding chemists to their university in an effort to not only raise the profile of the state but also to reinforce the university's research profile (James 1995). Such facilities coupled with outstanding mentoring attracted students from many different countries, who on their return home drew on their experiences and created similar research schools, thereby adding to the advance of chemistry.

In France, many regional centers for chemistry flourished during the eighteenth century, but it was between 1808 and 1810 that the government created fifteen designated Science Faculties across France in an attempt to decentralize from Paris (Nye 1986: 12–13). Ten remained within France after Napoleon's defeat in 1815. Each faculty was required to appoint a professor of physical sciences, although the final approval lay with the Ministry of Education in Paris. In addition to their teaching, professors had to administer and grade exams for the *baccalauréat* diploma. Other qualifications also formed part of the system: a licence required the additional step of an oral interview focused on a specialist topic; a doctorate required two dissertations, one on the specialist topic in the licence, and an oral examination; the *agrégation* diploma (a qualification required for teaching in *lycées* or university faculties) was achieved through a high-level examination. Further changes were made to the system in an effort to adapt it to the changing needs of the country: the *École Pratique des Hautes Études* was established in Paris in 1868; after 1870 the provincial Science Faculties were permitted to set up courses in applied science; in 1876 five Catholic universities were created; in 1896 universities were centered on the regional Science Faculties to boost research and counter German competitiveness (Fox and Weisz 1980: 23; Nye 1986: 20). As a result of the revised set of degrees in 1898 (administered by universities rather than the government), the doctoral degree benefited the provincial universities (Science Faculties) by attracting very many foreign students. However, the overwhelming majority of French students continued to study science disciplines in Paris rather than the provinces, at the *grandes écoles* (e.g. the *École Polytechnique* and the *École Normale Supérieure*), and research-oriented institutions that included the Collège de France, the *Muséum d'Histoire Naturelle* and the *Observatoire*). However, late in the century chemistry was to prove an exception, probably due to a few eminent chemists remaining in the provincial universities, such as Victor Grignard at Lyon and Nancy and Paul Sabatier at Toulouse.

In 1898 Grignard became laboratory director at the Lyon Faculties, and his collaboration with several colleagues led to interest in the use of magnesium in organometallic compounds and their role in synthesizing organic compounds (Nye 1986: 168). Grignard found that organomagnesium compounds were less flammable than the organozinc derivatives previously used, and chemists found them to have a wide range of uses for synthesizing organic compounds. Grignard was appointed professor of organic chemistry at the University of Nancy in 1912.

In 1877 Paul Sabatier was appointed to a physics post at Toulouse and collaborated with Jean-Baptiste Senderens on theories relating to the catalytic hydrogenation of hydrocarbons. After Senderens left Toulouse in 1905, Sabatier continued his research on catalysis. Appointed Dean of the Toulouse Science Faculty the same year, Sabatier brought about profound change by creating several specialist research institutes including the *Institut de Chimie* in 1906 (Nye 1986: 137–48). Sabatier and Grignard shared the 1912 Nobel Prize for Chemistry, confirming the influential role of French provincial institutions during the long nineteenth century and justifying the government policy of decentralization from Paris, largely unsuccessful though it was.

Higher education in the United States at the beginning of the nineteenth century, like Britain, focused on the classics, moral philosophy, and medicine. Up to 1850 chemistry was taught in medical schools as part of courses in *materia medica* and in a few colleges such as Rensselaer Polytechnic Institute (1824), but students with a serious interest in the sciences needed to study in Europe (usually Germany; Beardsley 1964: 1). Of the 10,000 Americans who graduated from German universities between 1850 and 1914, some 1,000 were advanced students in chemistry; many studied with Justus Liebig at Giessen and Munich, Robert Bunsen at Heidelberg, Friedrich Wöhler at Göttingen, or August Kekulé at Bonn. It was only during the second half of the century that American universities began to embrace chemistry as an academic discipline at both undergraduate and graduate levels, as a response to demands made initially by the geological surveys and agriculture and later by industry. A major factor in this expansion was the Morrill Act of 1862 that enabled each state of the union to use income from the sale of federal lands to endow its own state university focused on engineering and agriculture. By the end of the century many of these "land grant" universities included research, even doctoral programs, as an important part of their remit.

Leading roles were taken by the pre-Morrill Act privately endowed Sheffield Scientific School of Yale College (1847) and the Lawrence Scientific School of Harvard College (1847), followed by the Chandler Scientific School of Dartmouth College (1852), the University of Michigan (1817), and Massachusetts Institute of Technology (MIT; 1861), among many others. In 1847 Yale Corporation established the Department of Philosophy and the Arts that included the School of Applied Chemistry and appointed Benjamin Silliman Jr. as Professor of Chemistry as Applied to the Arts. With Silliman's later move to the University of Louisville (in 1849), the German influence continued with George Brush (previously a student at Giessen and Freiberg) taking up a post in mineralogy, and Samuel Johnson (previously a student at Munich) appointed an assistant in the chemistry laboratory (Dana 1920). In 1854 the department was renamed the Yale Scientific School and, with an endowment from Joseph Sheffield in 1860, was further renamed the Sheffield Scientific School and offered three-year undergraduate courses and a Ph.D. program (Beardsley 1964: 7).

The development of chemistry at Harvard College took a different route. Harvard received endowments that aided their expansion of scientific and technical education. The endowment of Abbott Lawrence (textile manufacturer and diplomat) in 1847 helped found the Lawrence School of Science.[4] The first Professor of Chemistry, Eben Horsford, organized the chemistry course at Harvard to follow the Giessen model, having studied under Liebig at Giessen during 1844–1847 (Munday 2000). The emphasis was on applied chemistry and Horsford drew on Liebig's vision linking chemistry with industrial development – a vision also reflected in Abbott Lawrence's endowment (Rezneck 1970). When Horsford left to work in industry in 1863, Oliver Wolcott Gibbs, who like Horsford had trained in Germany (in Giessen and Berlin) but also in France (Paris), was appointed to the Rumford Professorship of the Application of Science to the Useful Arts at Harvard College, and also took charge of the Lawrence chemistry laboratory. In 1871 the Lawrence School of Science was consolidated with Harvard College.

Johns Hopkins University in Baltimore (Maryland) occupies a unique position in American higher education. Founded in 1876 as a private research university through the benefaction of the wealthy merchant Johns Hopkins, the university adopted models represented by Giessen and Heidelberg Universities, linking teaching with research. Ira Remsen was appointed the first professor of chemistry, and used his experience of studying at Göttingen when designing laboratories and the research program (Figure 7.3). Between 1876 and 1913 Johns Hopkins awarded 202 Ph.D. degrees in chemistry. Overall, by 1900

FIGURE 7.3 Ira Remsen (center) and students in a chemistry laboratory, Johns Hopkins University, 1890. Photograph by JHU Sheridan Libraries/Gado/Getty Images.

the United States had a well-resourced higher education sector and was no longer entirely dependent on European institutions for the training of its future faculty.

Transmission of the chemical teaching–research laboratory to Japan followed a different route. The earliest Japanese students traveled to Britain at the end the Tokugawa Shogunate in 1863 and 1865. At the time, travel outside of Japan was illegal, but the districts of Chōshū and Satsuma, proponents of industrialization and modernization, defied the Shogunate to send these students abroad. In Britain, they attended University College London, where Alexander Williamson saw to both their pedagogical and personal needs. In 1867, four Japanese students traveled from London across the Atlantic to study at Rutgers College (now University) the newly established land-grant college in New Jersey.

More aggressive moves toward industrialization and Westernization in Japan would begin after the Meiji restoration of 1868. All Japanese students were now allowed to travel to the West to study chemistry, but the new government also realized that it would be more affordable, and they would have much more control over students' experience, if they brought foreigners to Japan to set up and direct chemical instruction. The Tokyo Kaisei Gakkō (which would become part of Tokyo University in 1877) hired a series of American and British chemistry instructors, largely drawn from the network established at UCL and Rutgers, to establish teaching laboratories. Although some conflicts emerged, the administrator Yoshinari Hatakeyama, one of the early Japanese students at the UCL who had great sympathy for the West (he converted to Christianity), but was also aware of his own origins in traditional Japan, skillfully mediated between the foreign teachers and the Japanese authorities. Students trained at the Kaisei Gakkō would study chemistry abroad and return to establish a more fully Japanese style of the teaching–research laboratory. At the Imperial College of Engineering, on the other hand, chemistry laboratories were also taught by British chemists, but remained more autonomous than those at the Kaisei Gakkō and the culture took on a more British air, with students eating Western meals and playing cricket and soccer (Kikuchi 2013). Regardless of the structure, by the late nineteenth century, Western-style chemical instruction had become firmly established in Japan.

ROLE OF FORMAL LECTURES AND PRACTICAL LABORATORY INSTRUCTION

Over the period there was a steady move to supplement formal lectures and demonstrations with practical laboratory instruction. Such an approach was not new but gained momentum in different countries at different times. Liebig and Bunsen in Germany were the models for this movement, and their international influence relied on their students taking the model back to their home country, although realization was often complicated by the additional cost of laboratories.

Several laboratory-based courses had been implemented in Europe early in the nineteenth century; this trend gained momentum in the second quarter of the century. The *École Polytechnique* in Paris had adopted a syllabus of experiments devised by Louis-Bernard Guyton de Morveau in 1806, and in the same year Friedrich Stromeyer began teaching a laboratory-based course at Göttingen University. From 1807 Thomas Thomson had started a practical chemistry course at the University of Edinburgh, and a similar course was provided at the University of Glasgow from 1817 when Thomson moved there. Edward Turner introduced a similar course at University College London in 1829, and the first teaching laboratory in the United States is credited to the Rensselaer School, in Troy (New York) in 1824 (Smeaton 1954: 232–3).

The greatest influence on the training of future chemists, however, was Liebig and his teaching laboratory at the University of Giessen between 1824 and 1852. Liebig's innovative approach had two stages. During the first, students were given practical exercises to develop a good understanding of chemical principles; in the second, students who had mastered the principles were assigned research projects as part of a team to advance Liebig's own larger research program. Beginning students and advanced workers all shared the same laboratory, as depicted in the well-known drawing of the laboratory (see Chapter 3, Figure 3.1). Liebig's close supervision allowed him to assess the progress of each student.

Liebig's students would spread his teaching–research model around the world. In Britain, Hofmann, Director of the Royal College of Chemistry in London from 1845 until 1865, and Sheridan Muspratt, founder of the Liverpool College of Chemistry in 1848, both drew on the Giessen model of laboratory-based learning. In France many leading chemists tried to implement the Liebig approach, although most did not succeed because of the cultural and political climate arising from the revolution. The principal exception was Adolphe Wurtz's privately operated laboratory at the Faculté de Médecine; Wurtz managed to gain government recognition, but only after twenty-three years (Rocke 2003: 112). In America, Eben Horsford returned in 1847 from a stint at Giessen to a professorial appointment at Harvard. With an endowment from the industrialist Abbott Lawrence laboratory-based teaching was established under Horsford's direction, but later proved largely unsuccessful due to the cultural and economic differences between Germany and the United States (Rossiter 1975: 87–8; Rocke 2003: 113).

Laboratory-based learning necessitated changes to laboratory design to create a functional and safe environment. Liebig's laboratory in Giessen and Bunsen's at Heidelberg were early models replicated across the world, but were later adapted and refined to accommodate the growing number of students, the growing specialist fields of chemistry such as organic chemistry, and the facilities for new instrumental techniques (Jackson 2011; Nawa 2014). Providing a safe working environment (with good ventilation and illumination) was an essential need. Supplies of water and gas (the latter for use with Bunsen burners) were required at the bench. Fume hoods and fume chambers allowed dangerous and obnoxious fumes to be vented to the outside. Hydrogen sulfide used in qualitative inorganic analysis (or group analysis as it was often called) was generated in a Kipp's apparatus, but because the gas is almost as toxic as hydrogen cyanide, the "Kipp's" was kept in a fume chamber to which students took their samples (Morris 2015: 108). Access to a fume chamber was also important when working with volatile organic chemicals with low flash points because of the risk of fire.[5] Through the nineteenth century special rooms (or laboratories), separate but adjoining the main laboratory, were added to accommodate sensitive functions such as weighing, gas analysis, electrochemical work, and spectroscopic analysis.

CHEMISTRY SPECIALTIES

The nineteenth century also saw chemistry divide into a number of different specialties. One of the most important was organic chemistry: the chemistry of compounds based on the element carbon that can also contain varying amounts of the elements hydrogen, oxygen, nitrogen, and occasionally other elements. It was especially Justus Liebig during the 1820s, 1830s, and 1840s at the University of Giessen who pioneered organic chemistry, developed techniques for quantitative analysis (including his *Kaliapparat*, or potash-bulb apparatus), and showed the link between chemistry and physiology and agriculture. At the Sorbonne and *École Polytechnique* in Paris, Jean-Baptiste Dumas was also substantively involved with the rise of organic chemistry during those same years. Inorganic chemistry, with its connections to mineralogy and crystallography, also developed apace during the nineteenth century. The rise of periodic systems of the elements spurred further search for new elements.

Physical chemistry applies the ideas and techniques of physics to chemical systems. Following important early work by figures such as Bunsen, Faraday, Hermann Kopp, and Henri Victor Regnault, it was advanced in the last third of the century by the work of Josiah Willard Gibbs, Wilhelm Ostwald, Jacobus Henricus van 't Hoff, and Svante Arrhenius who developed chemical thermodynamics, chemical kinetics, the theory of solutions, and the study of reaction equilibria. The 1880s saw the origins of chemical engineering – the thorough integration of chemistry and engineering (rather than the crude link between chemistry and mechanical engineering) – with the first course in that field conducted by the chemical consultant, George Davis at Manchester Technical College in 1888. Later in 1914 in the United States, Arthur Little at MIT proposed the important concept of "unit operation" to denote the core operations (such as crystallization, distillation, and heat transfer) that form essential components of processes on the industrial scale. Towards the end of the century biochemistry (biological chemistry or physiological chemistry) began to form a separate field of study. Early leaders in the field included Emil Fischer with his work on sugars and proteins at the universities of Erlangen, Würzburg and Berlin, and Frederick Gowland Hopkins with his work on vitamins at the University of Cambridge. These specialized fields also spawned a multiplicity of journals to disseminate the very latest research.

RESEARCH SCHOOLS AND SCHOOLS OF CHEMISTRY

During the nineteenth century, university chemistry departments that had embraced laboratory-based learning often established what amounted to a research school for their more advanced students. The phrase "chemist breeder" has been attributed to the research schools associated with Thomas Thomson at the University of Glasgow (from 1818) and Justus Liebig at the University of Giessen (from 1824; Morrell 1972). Key factors in successful research schools were good laboratory facilities, practical experience of laboratory work and perhaps the key, the supervisory and mentoring role of the professor.

Liebig had instigated a major change in the late 1830s when he began to organize groups of his students to focus on particular classes of organic compounds, assigning specific tasks, reviewing progress regularly, and shifting direction if required. While the students were allowed to theorize about their individual projects, it was Liebig who appreciated the broad overview and could integrate the individual research results into broader conclusions that might then prompt further research for the group (Holmes 1989b: 160, 162).

Later in the nineteenth century other schools of chemistry emerged under the leadership of outstanding chemists. Prominent among these research schools in Germany were: Robert Bunsen at University of Heidelberg; Friedrich Wöhler at the University of Göttingen; August Kekulé at the University of Bonn; Adolf von Baeyer at the University of Munich; and Emil Fischer at Universities of Erlangen, Würzburg, and Berlin. While these research schools operated under different approaches, the personality of the director proved vitally important in creating the best ethos for a steady stream of outstanding work (Rocke 1993: 114). The success of Baeyer's school was its size, research output, and longevity – between 1875 and 1915 some 560 chemists produced about 1,200 papers on organic chemistry (Morrell 1993: 106). Perhaps more important was that more than half of the 486 German and Austrian students went on to work in chemical industry or became commercial chemists during a period of major expansion of the German chemical industry (Fruton 1990: 158). A feature of Fischer's research groups was their focus on biochemical compounds such as sugars and carbohydrates. Fischer allocated specific tasks – synthesis and analysis of compounds – but students were instructed not to be influenced by theories or preconceived ideas, leaving their doctoral dissertations rather limited in academic scope (Fruton 1985: 317–19). The output of the German research schools (coupled with the *Technische Hochschulen*) was instrumental in establishing the German chemical industry as the world leader in the decades preceding World War I.

In Britain, Henry Roscoe at Manchester and William Henry Perkin Jr. at Manchester and the University of Oxford developed influential schools of chemistry. Drawing on his experiences working with Bunsen, Roscoe's approach had four component parts: the design and construction of laboratory accommodation that themselves became models for laboratories at other universities; the recruitment of outstanding chemists for teaching and research who later went on to other institutions; the stream of textbooks; and the constant flow of first-class research across the different specialisms of chemistry (Bud and Roberts 1984: 85–6). The "Owens model" was subsequently adopted (and adapted) by many universities in Britain and abroad.

Roscoe resigned in 1886 following his election as a member of parliament, and Perkin joined the Victoria University, Manchester (previously Owens College) as professor of organic chemistry in 1892. Perkin was greatly influenced by his German teachers, Johannes Wislicenus at Würzburg and Adolf von Baeyer at Munich in his expansion of the university's chemistry department, especially for the field of organic chemistry. He secured major external funding for new laboratories, including the Schorlemmer Memorial Laboratory (the first organic chemistry laboratory in Britain; Morrell 1993: 107–9). Perkin's work in Manchester (1892–1913) focused on expanding research opportunities in organic chemistry for those wishing to pursue careers in academia or industry (as Baeyer had done). At Oxford Perkin found a very different situation when he was appointed to the Waynflete chair in 1912; colleges had responsibility for teaching undergraduates and conducting research, there was only one university chemical laboratory, and organic chemistry was largely neglected. Perkin spurred the research output (although not always well received by university colleagues) and shifted its focus to work that would benefit industry (initially during wartime) in an effort to improve the prospects for the British organic chemical industry, then struggling against the domination of German industry. The Dyson Perrins Laboratory (1916–1922) increased the facilities for organic chemistry research; Perkin was a forceful and active administrator, persuading Oxford to adopt research degrees (Ben-David 1971: 119, 122).

PROFESSIONALIZATION OF CHEMISTS

The nineteenth century saw an increasing number of scientists and chemists taking a more prominent role in academia, industry, and government affairs, prompting the advance of professionalization. Societies were founded worldwide to represent chemists' interests, educational expectations for chemists were elevated, and journals proliferated to enable chemists to keep abreast of the latest developments in their field. Countries proceeded at their own pace and in line with their stage of development, although the societies in Britain often led the way and provided the early model.

By 1830 there was an emerging community of people in Britain who saw themselves as pursuing chemistry professionally and yet did not have an institution or society to represent their interests and advance the science of chemistry. There was general dissatisfaction among scientists with the foremost scientific society, The Royal Society, because of its exclusive and elitist character (Meadows 2004: 77–84). Two prominent British scientists, Charles Babbage (the inventor of early mechanical computers) and David Brewster (a natural philosopher in Edinburgh) were forthright in their condemnation of the Royal Society, and sought to create an alternative kind of organization. Following Brewster's suggestion, a meeting of British scientists was organized by the Yorkshire Philosophical Society in York during the summer of 1831 under the presidency of William Vernon Harcourt (Morrell and Thackray 1984: 48). Even though many prominent scientists did not attend, because they saw London as a more appropriate meeting place, the York meeting led to the formation of the British Association for the Advancement of Science (BAAS; Meadows 2004: 84). Overall, the aim was to meet annually and in a different city, following the approach of the German Society of Naturalists and Physicians formed ten years earlier. In this way the BAAS could address current major scientific issues as well as important national issues that science could influence. The public could also attend meetings and so engage more fully in scientific matters and meet prominent British and foreign scientists (Russell et al. 1977: 66–7). With newspapers and journals reporting the proceedings, meetings of the BAAS provided an annual showcase for British science.

While one of the BAAS's main aims was the unity of science, individual sciences were given a formal presence within its structure. In 1835 the individual sciences were designated as sections within the BAAS; chemistry became Section B. For each meeting, sections invited a prominent representative of the science to be president, and also asked a person from the host city to act as local secretary and organize speakers and visits to places of special interest. Presidents for Section B during the early years included John Dalton (1831, 1832, 1842), Thomas Thomson (1835, 1840), Michael Faraday (1837), and Thomas Graham (1839, 1844).

Britain was the first country to have a national chemical society, although chemistry was not the first science to have a national society. The Chemical Society of London was formed in 1841, and some historians have contended that its creation was probably the most important event in the history of chemical institutions (Russell et al. 1977: 68). However, before 1841 several local informal "societies" or gatherings of chemists had taken place (Averley 1986). For instance, in November 1824, a group calling itself The London Chemical Society was formally constituted, with George Birkbeck, a driving force in the Mechanics Institute movement, as president. Members were in the main enthusiastic young men; women were allowed to attend meetings but were excluded from membership. Unfortunately, this society was short-lived and may not have lasted into 1825 (Brock 1967: 138).

When the Chemical Society of London was formally constituted in 1841, its aim was to advance chemistry through regular meetings and the provision of a library, research laboratory, and museum, with the proceedings of its meetings published in its journal, initially entitled simply *Proceedings and Memoirs* of the Society. The Society's first president was Thomas Graham, professor of chemistry at University College London, and its first Foreign Member was the German chemist, Justus Liebig (Russell et al. 1977: 69). In 1848 the Society was granted its first charter and changed its name simply to the Chemical Society. It published the new *Quarterly Journal of the Chemical Society* with increased coverage of foreign developments in chemistry and its members were designated as fellows. The Society became the model for national chemical societies across the world (Brock 1993: 445).

The word "chemist" in British usage also designates pharmacists, and this ambiguity resulted in a number of ruptures and the formation of several different societies representing chemists. In July 1841 "chemists" who worked as apothecaries, druggists, and pharmacists formed themselves into an association, the Pharmaceutical Society of Great Britain, the same year that saw the formation of the Chemical Society.

Further professional fractures among chemists occurred from the 1870s, leading to two new professional bodies. With advances in analytical chemistry, many practitioners were designated analytical chemists (or consultants) during the first half of the nineteenth century. Much of their work concerned the detection of adulterations in foods and drugs and in the associated work of customs and excise (see Chapter 5). In 1874 (and following the report of the 1872 Select Committee on Adulteration of Food, Drink and Drugs), the Society of Public Analysts (SPA) was formed to protect the interests of its members against several conflicting issues that included their relationship with "academic chemists" as represented by the Chemical Society.

Analytical chemistry was not just confined to investigating adulterations but was also playing an increasing role in industries that included chemicals, brewing, mining, railways, metallurgy, explosives, agriculture, and gas production. As discussed in Chapter 5, the Alkali Inspectorate (formed in 1864) made use of analytical chemistry to ensure that the legal limits for noxious vapours were met, and also to provide scientific data for legal prosecutions should they prove necessary. While inspectors appointed by the Alkali Inspectorate were trained in analytical chemistry, many others acting as analytical consultants had medical qualifications. Chemists and those employing chemists became concerned about the professional standing and qualifications of those acting as analytical chemists. During 1877 a group of chemists outside the SPA began a secretive process to form a society separate from the Chemical Society. Many of those involved were prominent chemists working across a wide spectrum of chemical interests – academics, consultants, public analysts, medics, and veterinarians – but they were unified in their commitment to establishing a body to strengthen the professionalization of chemists. In October 1877 the Board of Trade incorporated the Institute of Chemistry (IC), with Edward Frankland as its first president. In 1885 the IC received its royal charter, and thenceforth, as the Royal Institute of Chemistry, the organization adopted discrete membership categories that necessitated specific qualifications and work experience. The Chemical Society, on the other hand, continued to concentrate on advancing the science of chemistry and publishing its specialist journals.

Another round of professional fragmentation occurred among chemists in 1881 when the Society of Chemical Industry was founded. Before 1880 there were societies in Tyneside (northeast England) and Lancashire (northwest England), both areas where the chemical industry dominated the industrial landscape. Tyneside had two societies – the Newcastle Chemical Society and the Tyne Chemical Society. Lancashire had two societies. The Faraday Club, formed in 1875, brought together chemists in Widnes and St. Helens, both major centers for chemical manufacture. In 1879 a South Lancashire Chemical Society was proposed; through a series of meetings it became the Society of Chemical Industry to embrace the large number of chemists working in the chemical industry, although several of the prospective members had sought the name "Society of Chemical Engineers." At a meeting held in the rooms of the Chemical Society on April 4, 1881 the Society of Chemical Industry (SCI) was formally constituted, with Henry Roscoe of Owens College, Manchester elected as its first president. It would have local sections and an annual meeting held in a different city each year. There was an active group of SCI members in New York; after informal meetings and discussions with the SCI in London the first meeting of the official New York section was held in 1894. The SCI's annual meeting was held in New York in 1904 and 1912, reflecting the society's strong outreach to embrace its membership in the United States (1931: 17).

There were several smaller societies in Britain that brought together those working in more specialist areas of chemistry including its practical applications. The Society of Dyers and Colourists was founded in Bradford (West Yorkshire) in 1884 and brought together all those engaged with dyestuffs, dyeing, and colors in general. The Society published the *Journal of the Society of Dyers and Colourists* from 1884. The Faraday Society was founded in 1903 following the spectacular advances in physical chemistry during the 1860s, 1870s, and 1880s alluded to earlier in this chapter. Ostwald's regular visits to Britain and his meetings with British chemists did much to promote physical chemistry and the founding of the Society. The Society published the *Faraday Transactions* from 1905. In 1911 the Biochemical Club was founded, and the following year it took over publication of the *Biochemical Journal*. The name of the Society was changed to the Biochemical Society in 1913.

The United States saw a later but similar pattern in the formation of its chemical societies as Britain. The American Association for the Advancement of Science (AAAS) was formed in 1848 in Philadelphia. The aims of AAAS were similar to the BAAS: focus on the unity of science and on its promotion, elevate the profession of science, and raise funds to advance research. It had a similar structure to the BAAS with each specialist science (including chemistry) allocated a section. Its meetings between 1861 and 1865 were suspended due to the American Civil War, but resumed in New York City in 1866.

Two important multiscience organizations were also formed during the nineteenth century. In 1817 the Academy of Natural Sciences of Philadelphia became the pre-eminent organization for natural sciences in the United States with its outstanding collections and research. In 1863, Congress established the National Academy of Sciences that operated in a similar manner to the Royal Society in London; its membership was by election and based on the standing of published works.

By the early 1870s the influence of chemistry across the United States prompted a group of chemists to consider forming a society to advance "pure" as well as applied chemistry. Initial discussions took place after the 1873 AAAS meeting in Portland (Maine) and led to the formation of the American Chemical Society (ACS) in New York in 1876. The ACS struggled for many years with a declining membership as it was seen more as a "New York club" rather than a society for chemists across the whole of the United States (Thackray et al. 1985: 22–3). A further review in 1889 focused attention on the structure of Britain's Society of Chemical Industry with its local sections spread out across the regions of the country. In 1891 the new organization of ACS adopted a revised structure incorporating sections: the constitution of the original New York ACS was changed in 1890 and the New York chemists became a section of the revamped ACS.

The first society in Germany to foster an interest in chemistry was the Society of German Naturalists and Physicians (*Gesellschaft Deutscher Naturforscher und Ärzte*), founded in Leipzig in 1822. The explorer and naturalist Alexander von Humboldt was the first secretary, and at the Society's seventh meeting held in Berlin Humboldt drew up arrangements that became a model for its meetings (and would influence the BAAS and the AAAS). The proceedings were divided into seven specialist sections whose meetings took place before and after the main (or plenary) sessions. To engage with a wider audience, some meetings were also open to the public. As in Britain and the United States, a German society devoted to all aspects of chemistry came later. It was not until 1867 that the German Chemical Society (originally the *Deutsche Chemische Gesellschaft zu Berlin*) was founded, and like most other national chemical societies was modeled on the British Chemical Society. The Society's journal, *Berichte der Deutschen Chemischen Gesellschaft*, was first published in 1867. Adolf Baeyer was a leading figure in the foundation of the Society, but the most influential founder and the Society's first president was August Wilhelm Hofmann, who on his return to Germany from Britain in 1865 had been appointed professor of chemistry at the University of Berlin. By 1892 when Emil Fischer was president, the German Chemical Society was the world's largest chemical society, and its *Berichte* the most prestigious chemical journal (Johnson 2017).

The French Chemical Society (*Société Chimique de Paris*, renamed in 1906 the *Société Chimique de France*) was founded by three chemistry students in 1857; in the following year the outstanding French chemist Adolphe Wurtz became a member and transformed the society, modeling it on the Chemical Society in London (Rocke 2000a: 219). During the period to 1915 many other countries across the world founded national chemical societies: Russia (1868), Japan (1878), Denmark (1879), Sweden (1883), Norway (1893), the Netherlands (1903), Hungary (1907), and Italy (1909) (Nielsen and Strbánová 2008).

The professionalization of chemistry during the nineteenth century was also promoted by the growth in academic journals. Many of these journals were associated with the new chemical societies, but many others reflected the commercial opportunities for wider communication of chemical knowledge, as research output increased and chemistry fragmented into more specialties. In the early decades of the nineteenth century journals were not always devoted exclusively to chemistry, but covered several sciences to reflect "natural philosophy." In 1815 chemistry was covered in several journals, including *The Philosophical Transactions* of the Royal Society of London, and *Annales de chimie*, first published in 1789 in France, although in 1815 its name changed to *Annales de chimie et de physique*.[6] Like the *Annales de chimie*, the German *Annalen der Chemie* occupied a prestigious place among chemistry journal publications. Justus Liebig acquired the *Magazin der Pharmacie* in 1831 and renamed it *Annalen der Pharmacie* the following year. As editor, Liebig used the journal partly to communicate the research output of his Giessen laboratory and to engage in controversial debates. In 1840 the title was changed to *Annalen der Chemie und Pharmacie* and after Liebig's death in 1873, *Justus Liebig's Annalen der Chemie* (Brock 1993: 446–7).

The first journal in Britain exclusively devoted to chemistry was probably *The Chemist*, edited by Thomas Hodgkin, but it had a short life with only two editions, in 1824 and 1825. Between 1840 and 1858 *Chemist; or Reporter of Chemical Discoveries and Improvements* was published with an emphasis on medical and pharmaceutical topics (Brock 1993: 437). Publication of William Crookes' *Chemical News* in 1859 marked an important stage in the expansion of journals devoted to chemistry. Crookes was an outstanding chemist and used his knowledge of chemistry and his wide network of scientific contacts to gather sufficient topical news and information to sustain a weekly publication (Brock 1993: 454–64). As mentioned earlier, the Chemical Society of London had its own journal.

In America, chemistry was initially covered in general science journals. The first was the *American Journal of Science* edited by Benjamin Silliman Sr. and published from 1818, which focused on the latest research. Then in 1826 the *Journal of the Franklin Institute* appeared, focusing on US patents. The first chemistry journal proper, *The American Chemist*, appeared from 1870, edited by Charles Chandler (Columbia College of Mines) and William Chandler (Columbia School of Mines and later Lehigh University) with an emphasis on applied chemistry (Thackray et al. 1985: 176). With the need to communicate the rapid advances in chemistry, the *Journal of the American Chemical Society* appeared from 1879, while the *American Chemical Journal*, edited by Ira Remsen (at Johns Hopkins University) was started the same year to stimulate basic research and merged in 1913 with the *Journal of the American Chemical Society* (Thackray et al. 1985: 177). In the 1890s the former concentrated on organic chemistry while the latter provided a detailed summary of American research. Later specialist journals began to appear, including: *Journal of Physical Chemistry* (1896), *Proceedings of the American Electro-Chemical Society* (1902), and *Biological Chemistry* (1906). These journals severed "the last bond of dependency which tied it [American chemistry] to Europe" (Beardsley 1964: 42).

CONCLUSION

By 1914, chemistry had emerged as the first modern and most readily applied science. Its position had been attained through the transforming inclusion of science in school and especially university curricula. Laboratory-based learning became a universally adopted technique for learning chemistry, initially based on the model developed by Liebig, but then replicated across many other countries of the world (invariably in a modified form) by students who had studied with Liebig. The large body of chemists worldwide sought professional recognition through national chemical societies, often based on the British model but adapting to the priorities and demands of individual nations. With the rapid pace of chemical research, societies saw a role for communicating the latest knowledge to their members through publication of journals. Chemistry fragmented into specialties such as electrochemistry, physical chemistry, and organic chemistry that spawned new journals for those working in the specific field.

The adoption of innovative laboratory-based learning led many German chemists (under the initial leadership of Liebig and Bunsen) to develop university research schools that transformed the conduct of research activity and accelerated chemical advance. These schools were replicated in Britain and the United States and strengthened their research outputs. The scale of Germany's chemical manpower and management of the knowledge from their research schools gave German chemical companies (together with their own research) such a huge advantage that the country's chemical industry became the most powerful commercial force in the world by 1870. Research on the conversion of nitrogen into ammonia for fertilizers with which to expand food production had led to the Haber–Bosch process (Chapter 4 and Volume 6 in this series.).

CHAPTER EIGHT

Art and Representation: The Rise of the "Mad Scientist"

JOACHIM SCHUMMER

INTRODUCTION

Compared to traditional themes of love, crime, and power, science plays only a small role in nineteenth-century literature. That is not surprising if we consider that science was then hardly comparable in size and influence to our times. And yet, science is clearly visible in the literature, and among all the disciplines chemistry plays a very prominent role.

Chemistry's visibility is not so much because writers and artists were acquainted with the field and employed its latest concepts and theories in their work. A rare exception was Johann Wolfgang von Goethe, an enthusiast and connoisseur of chemistry who based his novel about changing romantic relationships (*Wahlverwandschaften*, 1809, English: *Elective Affinities*) on contemporary ideas of chemical affinities. Instead, the dominating and overarching literary theme for the entire century in most Western countries, much to the dismay of scientists since then, was the "mad scientist," who as a rule is a chemist or a physician performing chemical experiments.

Up to the 1990s the depiction of science in literature had received very little attention from scholars of literature studies because of barriers between the humanities and the sciences, including their historiography, and the monolingual or national focus of the humanities. Unlike alchemy (or some allegorical version of it), which through C.G. Jung's psychoanalytical approach became a favorite subject, chemistry has remained a stepchild of literature studies, despite its powerful role in shaping the overall public image of science through the "mad scientist" trope.[1]

Although the term "mad scientist" was used first in the twentieth century for movie adaptions of literary themes from the previous century, the trope was created by nineteenth-century writers, who drew on earlier tropes, in particular the "mad alchemist," a literary figure developed in satire since the fourteenth century, and the sixteenth-century Faust legend. The historical background allows us to recognize the emergence of the "mad scientist" with this alchemical legacy as a stepwise transformation from the late eighteenth century onwards. The following three sections describe this genealogy from the "mad alchemist" to the standard "mad scientist" and highlight its most famous version, Mary Shelley's *Frankenstein*. In the standard form, the scientific madness consists largely of the obsession with some laboratory work on material perfection combined with hubris, moral naivety, and social seclusion.

Once the standard trope was established, writers developed variations on it by modifying "scientific madness" to include apparent schizophrenia, as for instance in *Dr Jekyll and Mr Hyde*, and various forms of moral disorder, from perverted means and goals in the name of science to criminal intent. At the end of the century the "mad scientist" was also transformed into more benign figures, such as the "benevolent but absent-minded scientist," and the heroic detective Sherlock Holmes.

The "mad scientist" trope, including a variety of fictional capabilities attributed to chemistry, dominated nineteenth-century literature such that the actual societal impacts of chemistry, including industrialization and pollution, remained almost invisible. In contrast, visual representations of chemists and chemistry, which were mostly commissioned by chemists to demonstrate achievement and social status, lack any reference to the trope. The "mad scientist" began to dominate the visual representation of chemistry only in the early twentieth chemistry with film adaptions of literary classics, which could draw on the iconographic repertoire of the "mad alchemist." Finally, I provide an overall account of the literary image of chemistry that goes beyond the "mad scientist" and suggest why that stereotypical image was so negative.

PRECURSORS TO THE "MAD SCIENTIST": THE "MAD ALCHEMIST" AND THE FAUST LEGEND

The concept of the "mad scientist" that would dominate the nineteenth-century literary depiction of science in general and of chemistry in particular goes back to the late-medieval "mad alchemist," a literary product from early debates about alchemy (Schummer 2006). In the early fourteenth century, certain European kings, particularly in France and England, produced counterfeit money to finance their wars and to pay their tithe to the Pope, which resulted in a papal bull against alchemy in 1317, denouncing it as fraudulent (Ganzenmüller 1938; Ogrinc 1980; Obrist 1996). Many kings who officially forbade alchemy nonetheless continued financing it in secret for centuries, and others would do as well.

Numerous authors, particularly those who supported the papal supremacy, reinforced the critique of alchemy (Read 1947; Linden 1996; Meakin 1995). In the fourteenth century, they included influential philosophers, such as Nicole Oresme; inquisitors who, like Nicolaus Eymericus, saw in alchemical practice demons or devils at work; and newly emerging fiction writers such as Dante Alighieri who put alchemists in the deepest hell in his *Divine Comedy* (1310–1321, Inferno: Canto XXIX). Francesco Petrarch was the first to vividly describe how a miserable seeker of gold-making ruins his life (*Remedies for Fortune Fair and Foul*, 1353–1366, chapter 111 "On Alchemy"). Soon Geoffrey Chaucer would turn the alchemist into a wretched figure of satire in his *Canterbury Tales* (*ca*. 1390, "Canon's Yeoman's Tale"), starting a popular literary tradition, including for instance Sebastian Brant's *Ship of Fools* (1494, chapter 102), Erasmus of Rotterdam's *Colloquies* (1518, "Beggar Talks" and "Alchemy"), Reginald Scott's *The Discoverie of Witchcraft* (1584, 14th book), and Ben Jonson's *The Alchemist* (1610).

In its fully developed version, the "mad alchemist" consists of two figures or phases, the "miserable seeker" and the "tempter" or "cheating alchemist." The "tempter" infects the "miserable seeker" with the obsession of gold-making by some trickery and promises of richness. The "seeker" receives or buys substances, equipment, and technical guidance from the "tempter" and secludes himself in a laboratory where he works day and night. He ruins his health from the smoke and other poisons, isolates himself from his social environment, loses his public reputation, spends all his money on the work, and brings poverty and shame on his family (if he has one). As a drug addict can become a drug dealer, so also the "miserable seeker" in a second phase may turn into a "tempter." He then infects other victims whom he sucks dry to finance his own obsession. If the "tempter" himself is not an addict, he is usually said to come from abroad and bears various devilish attributes.

The extraordinary literary success of the "mad alchemist" theme, a truly medieval invention without ancient models, was probably due to its usefulness in satire and its general moral effect. The "miserable seeker" ruins himself because of both his stupidity and greed, two widespread follies against which many authors wrote. In Chaucer's story, a yeoman, a canon, and a priest are subject to satire. Erasmus made fun of the stupidity and greed of priests and academic scholars. Jonson ridiculed the entire nobility and middle class. In Scott's story, a greedy king is fooled.

During the eighteenth century the popularity of the "mad scientist" temporarily faded. On one hand, other literary tropes, such as the "gamester," emerged that could serve the same purpose. On the other hand, alchemy probably became more popular in Europe, where circles devoted to various kinds of obscurantism greatly flourished during the so-called time of Enlightenment.

After Gutenberg's revolution in printing technology around 1450, many of the books mentioned went into print and were illustrated with woodcuts, including the best-selling books by Petrarch and Brant, most likely illustrated by Albrecht Dürer and Hans Weiditz the Younger, respectively.[2] Their visual representations of an alchemist working in the laboratory became highly influential for the subsequent imagery of the "mad scientist" in nineteenth-century literature and twentieth-century movies. Weiditz depicted in 1519 the laboratory overstuffed with partly broken alchemical apparatus (alembics, retorts, various vessels, bellows), and the alchemist in worn-out clothes obsessively working at the smoking furnace, while his assistant stands in the background scratching his head. In Dürer's woodcut from 1494, the "mad alchemist" is of higher social status and still at an early state of his obsession, because he is performing a distillation at the furnace under the guidance of the "tempter." It was obviously modeled on a popular image circulating at the time (around 1480), in which the alchemist is instructed by the Antichrist with the devil acting in the background.

Through Pieter Bruegel the Elder's famous painting *The Alchemist* (1558), which depicts both the final state of the alchemical obsession by a farmer and the entry of his family into an almshouse, the "mad alchemist" motif moved to Flemish genre painting. In the seventeenth century David Teniers the Younger (1610–1690) and many of his colleagues produced dozens of satirical alchemist paintings, which they sold all over Europe. Moreover, they merged the then famous motif of quack medicine, a man gazing at a flask filled with some colored liquid representing uroscopy, with the mad alchemist motif, because both the quack and the alchemist were considered cheaters. Strangely enough, late-nineteenth-century chemists would adopt exactly that posture of gazing-at-a-flask as the emblematic image of their science (see below).

In the sixteenth century, Johann Spies created the influential literary character Faust in *Historia von D. Johann Fausten* (1587), which Christopher Marlowe almost immediately popularized in the English-speaking world (*Dr. Faustus*, 1589). Faust is an alchemist, but also a healer, astrologer, and necromancer. He is not just tempted by the devil, but makes an explicit deal with him in order to increase his technological power, a feature of the story that makes the Faust legend clearly derived from the "mad alchemist." In the early nineteenth century, Goethe would reimagine the Faust story into a romantic tragedy of a scholar longing for wisdom beyond human reach, in stark contrast to the "mad scientist" of the nineteenth century.

THE STANDARD "MAD SCIENTIST"

At the turn of the nineteenth century, writers of many countries revived the trope of the "mad alchemist" and transformed it into the standard form of the "mad scientist." From that trope, various other types were later derived by variations in "madness," to be dealt with in subsequent sections. Like his precursor, the standard "mad scientist" is obsessed with some chemical project, secludes himself in the laboratory, works day and night, and becomes socially isolated. As a rule he is a male chemist, amateur or professional, or a physician turned to chemical research, and his laboratory is largely equipped with the stuff of the "mad alchemist." His goal might still be gold-making, but it could also be creating diamonds, elixirs for longevity, antidotes to poisons, new medicines, artificial life, or other things of public fascination; generally forms of material perfection according to widely shared values. He might still be guided by selfish motives, but usually he works "in the name of science," seeks scientific fame for his accomplishments, and most often does it for the betterment of humanity. Unlike the "mad alchemist," the "mad scientist" is typically successful in his scientific deeds. Thus, in contrast to medieval and early modern writers, modern authors attributed tremendous creative capacity and power to chemistry.

Thus far, the "mad scientist" appears to be an omnipotent benefactor to humanity by successfully achieving goals of material perfection. However, in all standard "mad scientist" stories the desired goals, once achieved, turn into evil by some unforeseen consequences or by accident. The general lesson is that scientists are powerful in their deeds but disastrous in the consequences of their actions, because they lack a broader understanding of the social and moral context in which they act. In other words, their scientific aspiration for improvement is morally naive.

I will first discuss some standard examples that illustrate the transformation from the mad alchemist, then highlight the most famous example, Mary Shelley's *Frankenstein*, before I deal with the modifications.

Already in 1799, William Godwin, a political philosopher and novelist, and the father of Mary Shelley, published *St. Leon: A Tale of the Sixteenth Century*. After Count Reginald de St. Leon learns the secrets of both gold-making and an elixir of rejuvenation from a foreigner, his life turns into despair. Whenever he makes gold, he is persecuted by officials, to whom his sudden wealth is suspect, even though he tries to spend it for the benefit of the poor. He is forced to leave his family and wanders aimlessly through Europe, letting his children believe that he has died in order to restore their reputation. At times he is put into a dungeon to practice gold-making for others. And when after decades he first uses his elixir of youth, he almost destroys the love of his son.

Whereas Godwin set his "mad scientist" plot in the sixteenth century, as Robert Browning also did in his play *Paracelsus* (1835), others openly portrayed contemporary chemistry in the disguise of alchemy. For instance, in Honoré de Balzac's (1799–1850) "mad scientist" novel *La Recherche de l'absolu* (1834, English translation as *The Quest of the Absolute*, *The Alkahest*, or *The Philosopher's Stone*), Claes is a former student of Lavoisier, who ruins his wealth, health, reputation, and family with his only partly successful obsession with synthesizing gold, diamonds, and organic substances.[3] Despite the use of various details from modern chemistry, the plot is still very close to the medieval "mad alchemist" stories, as are, for instance, Friedrich Halm's play *Der Adept* (1836, *The Adept*) and Wilkie Collins' *Jezebel's Daughter* (1880), albeit put into different narrative styles.

Thematic variations were introduced either by new and surprising endings or by novel "scientific" motives. Successful gold-makers not only face the envy of their neighbors and their own moral corruption, but also initiate the process of economic inflation that makes their gold worthless, as in Wieland's *Der Stein der Weisen* (1786–1789, *The Philosopher's Stone*) and Edgar Allan Poe's short story "Von Kempelen and His Discovery" (1840). Diamond-making replaced gold-making, as in Jean Paul's satirical novel *Der Komet oder Nikolaus Marggraf* (1820–1822, *The Comet or Nikolaus Marggraf*) or in H.G. Wells' short story "The Diamond Maker" (1894), where after fifteen years of obsessive work and on the edge of starving an amateur chemist is eventually successful in his experiments, but runs into trouble with the police when trying to sell his diamonds.

The traditional literary use of elixirs of life was superseded by chemical research with poisons and antidotes, usually performed by doctors who were considered to experiment with life and death, a form of "playing God." The American author Nathaniel Hawthorne set the stage with one of his most famous "mad scientist" short stories, "Rappaccini's Daughter" (1844), where of a doctor it is said that "He would sacrifice human life, his own among the rest, or whatever else was dearest to him, for the sake of adding so much as a grain of mustard seed to the great heap of his accumulated knowledge." Dr. Rappaccini has fed his daughter since early childhood with poison to make her strong and resistant against the rigors of the world, but then she happens to come in contact with an antidote and dies. In Wilkie Collins' novel *Heart and Science. A Story of the Present Time* (1883), Dr. Benjulia's obsessive research with poisons and antidotes, for the betterment of science, kills not only numerous test animals and leads him to poverty; his posthumous laboratory remains are also abused by others for committing murder.

Although the creation of life would much later become the most popular "mad scientist" topic through movie adaptions of Mary Shelley's *Frankenstein* (see below), it does not play much of a role in the nineteenth century. Another famous literary use of the "creation of life" theme is in Goethe's *Faust II* (1832), but that is not a typical "mad scientist" story. Faust's former famulus, Wagner, who has become a doctor of science and concocts a homunculus in his chemical laboratory, is portrayed as an idiot. Once alive, the homunculus leaves his surprised creator back in his laboratory and joins Faust and Mephistopheles on their adventure trips. In H.G. Wells' *The Island of Doctor Moreau* (1896), a physician creates new animals by surgery to improve their natural evolution, but that does not involve any chemistry. The first "mad scientist" story that features the chemical creation of simple life forms is perhaps "Microcosmic God" (1941) by US science-fiction author Theodore Sturgeon.

During the nineteenth century the "mad scientist/alchemist" was featured not only in novels, short stories, and plays but also in other literary forms and media, including fairy tales – e.g. "Das Wasser des Lebens" (1812–1815, "The Elixir of Life") by the Grimm Brothers; "Der Stein der Weisen" (*ca.* 1835, "The Philosopher's Stone") by Hans Christian Andersen – and even numerous operas (Griffel 1990).

MARY SHELLEY'S *FRANKENSTEIN*

Mary Shelley's *Frankenstein, or the Modern Prometheus* (1818/1831) has become the most famous literary example of the "mad scientist." When scientists create something that is otherwise known only from natural sources, today's media frequently comment on it with the prefix "Franken," such as in "Franken-food" or "Franken-bacteria." The media thereby seem to refer to a simple plot – a "mad scientist" creates a human-like being who runs amok – whereas the novel itself is quite unusual in two regards.

First, in ethical terms *Frankenstein* is probably the most eloquent and deliberative "mad scientist" story that has ever been written. Shelley organized the book with alternating moral monologues by Victor Frankenstein and his creature, who each reflect on their situation, what they have done, why they did it, and how they should be morally judged, including strong self-accusations. Second, it includes the most radical criticism of science one can think of. Rather than just pointing out a certain failure or some scientific misconduct by a man who has lost his senses, it suggests that the failure is an inevitable result of science in general, of scientific curiosity, no matter how high its moral standards are. That becomes clear if we look closer at Victor's career in science.

In the first chapters Shelley has Victor narrate in retrospect his childhood, adolescence, and studies at the University of Ingolstadt. The narrative recapitulates the history of Western science, such that Victor's years of age each correspond to the previous centuries (Schummer 2008). At thirteen he becomes interested in science, "the birth of that passion which afterwards ruled my destiny" (Shelley 1831), such as the Latin world, in the thirteenth century, became interested in natural philosophy after the first texts of Greek science had been translated from Arabic to Latin. Between thirteen and fifteen he reads texts by Albert the Great, Agrippa of Nettesheim, and Paracelsus, authors of natural philosophy and alchemy from the thirteenth to sixteenth centuries. "Under the guidance of my new preceptors I entered with the greatest diligence into the search of the philosopher's stone and the elixir of life" in order to "banish disease from the human frame and render man invulnerable to any but a violent death" (27).

Next, Victor is confronted by entirely new theories presented by "a man of great research in natural philosophy," probably inspired by Descartes, who visits his family and explains electric phenomena. He falls into a temporary period of (Cartesian) skepticism, remarking that "It seemed to me as if nothing would or could ever be known" (28), after which he, at seventeen, focuses on mathematical studies because they are "built upon secure foundations," such as the rising mechanical philosophy in the seventeenth century including Newton. Interestingly, Victor describes that period of mechanical studies as "the last effort made by the spirit of preservation to avert the storm that was even then hanging in the stars and ready to envelop me." However, "Destiny was too potent, and her immutable laws had decreed my utter and terrible destruction" (27).

When he is eighteen, an age that corresponds to the eighteenth century, Victor enrolls in the University of Ingolstadt to study natural philosophy, where the subject is already fully dominated by chemists teaching a new doctrine. He first meets Professor Kempe, who is still busy criticizing the old doctrines, reflecting the debates of the Chemical Revolution in the late eighteenth century. Then he listens to a lecture by Professor Waldman, who develops great visions that make Victor an adept of the New Chemistry. Modern scientists, says Waldman, "penetrate into the recesses of nature and show how she works in her hiding-places. ... They have acquired new and almost unlimited powers" (34). For Victor, in retrospect, those were "the words of the fate – enounced to destroy me" because "From this day natural philosophy, and particularly chemistry, in the most comprehensive sense of the term, became nearly my sole occupation" (36).

Victor soon works in Professor Waldman's laboratory, making a name for himself in the invention and improvement of chemical instruments, which corresponds to historical efforts at gravimetric and electrochemical instrumentation in the late eighteenth and early nineteenth century. At the age of twenty, when the nineteenth-century novel turns into a futuristic vision of the twentieth century, Victor begins researching the causes of life. With the obsession and self-sacrifice of a "mad alchemist," he tries to revive a dead body that he has constructed from parts of human corpses. Details are missing, but again Shelley presents the final step as a natural consequence of the previous science. The crucial idea was "so simple ... I was surprised that among so many men of genius who had directed their inquiries towards the same science, that I alone should be reserved to discover so astonishing a secret" (38).

Through Victor's biography, Shelley in effect narrates the history of science, from the Middle Ages to the future, as a necessary and predestined process towards the final catastrophe. Good intentions cannot prevent that disaster, but even foster its occurrence. As Victor reflects on his life: "I had begun life with benevolent intentions and thirsted for the moment when I should put them in practice and make myself useful to my fellow beings" (74). Because neither his good intentions nor the predestination of science can excuse Frankenstein's guilt, the only option would be to abstain from science. And thus Shelley has the dying Victor say: "Seek happiness in tranquility and avoid ambition, even if it be only the apparently innocent one of distinguishing yourself in science and discoveries" (196). In later adaptations of *Frankenstein* – beginning with the 1823 play *Presumption; or the Fate of Frankenstein*, to the many Hollywood movie productions in the twentieth century – both the sophisticated moral discussion and the general theme of the tragic predestination of science were removed.

Once the "mad scientist" had become an established figure, any kind of madness would suffice for the purpose of entertainment. The following five sections discuss the five most important types of madness, each of which was established in the nineteenth century as a distinctive variation on the original form of the "mad scientist."

PSYCHOLOGICAL TWIST

British "mad scientist" stories frequently come with a psychological twist that goes beyond the madness of the standard form. That twist has fascinated scholars of literature studies who have suggested various Freudian-style interpretations, but a simpler interpretation becomes obvious if we compare a few examples.

Frankenstein, my first example, stands out because it is composed of alternating monologues by Victor and his creature. Both reflect on their own individual but related situation with the same kind of sophisticated moral reasoning. At times it is even difficult to distinguish the two voices. Scholars have argued for some time that Shelley intended Victor's creature to be his alter ego, a motif that she borrowed from British gothic novels of the late eighteenth century (Tropp 1976: 37). However, the alter ego is not just an imagined part of Victor's soul. Instead, Victor creates the creature "in his own liking," or better, what he considered it to be, i.e. a being with the physiological and intellectual capacities of a human. However, what he has forgotten to add to his creation, and what the creature deeply complains about and what makes him eventually run amok, are the emotional and social sides of human beings. Victor is successful in the physiological and intellectual parts of his experiment, but he profoundly fails because his creature possesses an incomplete soul.

The second and most famous example of a "mad scientist" with a psychological twist is Robert Louis Stevenson's novella *Strange Case of Dr Jekyll and Mr Hyde* (1886). Dr. Jekyll is a physician who has turned to chemical research on psychiatric drugs. He develops both a psychological theory and two related drugs, the first drug to separate the good and evil forces of the soul and the second one to reunite them. Because he is unhappy with his own temper, which at times he is unable to control, he tests his theory and the drugs at first on himself. That temporarily turns him into the violently aggressive Mr. Hyde, who commits murder and other violent crimes. When the aggressive personality begins to dominate him and the second drug runs out because of lack of supply of chemicals, he commits suicide while in the temporary state of Dr. Jekyll to prevent the world from suffering more evil from Mr. Hyde. As with Victor Frankenstein, Dr. Jekyll fails because his science, while providing effective chemical tools, is based on an incomplete understanding of the human psyche.

The third, psychologically more twisted example is Charles Dickens' fourth Christmas story, *The Haunted Man and the Ghost's Bargain* (1848), about the chemistry professor Mr. Redlaw. The poor man is haunted by memories of grief and pain and makes a Faustian bargain with a ghost, who is actually his alter ego. He would lose all his painful memories if, in return, the entire emotional side of his memory is deleted and the same mental condition spreads to anyone whom he meets. As a result, he is soon surrounded by people who lack any social and emotional capacities that would rely on their memory, and Redlaw becomes even unhappier. Although Dickens' "thought experiment" includes a mysterious Faustian bargain (with oneself) rather than scientific manipulations of the human soul, the story follows a similar line as the other two examples.

As a rule, the "mad scientist" stories with a psychological twist are focused on creating or tampering with the human psyche, rather than producing gold, diamonds, elixirs of life, or other material entities. They tell the same story of scientific hubris and failure by lack of understanding and circumspection, despite the powerful scientific tools for manipulation in their hands, as in the standard "mad scientist" stories. Thus, the psychological twist is not a variation of the original "scientific madness," as in the subsequent examples, but it derives from the specific goal of tampering with the human psyche.

MORAL PERVERSION

Both the standard "mad scientist" and the version with a psychological twist seek some kind of perfection according to values largely shared by society, but they fail for lack of circumspection. In ethical terms, even though they might have had good intentions, they do wrong because their actions have adverse effects that they naively did not foresee. Several nineteenth-century authors, particularly from France, modified the "mad scientist" by varying the ethical theme, such that the "scientific madness" turns into moral perversion rather than moral naivety.

In one version of this moral perversion, achieving an acceptable goal requires methods that are morally wrong. For example, in Alexandre Dumas' fictional autobiography of Cagliostro (*Joseph Balsamo: Mémoires d'un médicin*, 1844–1846), to prepare an "elixir of life," Cagliostro's master of alchemy, Althotas, demands the last drops of blood from a child, for which the child must of course be murdered. The prolongation of one life requires the death of another, an act unscrupulously approved by the "mad scientist."

A second version of moral perversion involves immoral research goals. The "mad scientist" is mad because he has lost any sense of cultural reality and moral values, and ruthlessly pursues his strange goals. His activity needs to be distinguished from mere crime (see below) by pointing out that he allegedly acts "in the name of science" or even for the benefit of humanity. This type of "mad scientist" first appeared in nineteenth-century literature, and would later become a frequent fixture in Hollywood movies of the mid-twentieth century.

For example, the French pioneer of science fiction, Jules Verne, frequently described as a utopian, developed a particularly salient form of the moral perversion of "scientific" goals. In his short story *Une fantaisie du docteur Ox* (1872, *Dr. Ox's Experiment*) the main character, "an able chemist as well as an ingenious physiologist," visits a rural town where people live a peaceful, calm, and happy life. Dr. Ox convinces them that they would benefit from a gas lighting system fueled by hydrogen. And so he builds a factory for the electrolytic dissociation of water into hydrogen and oxygen.

However, Dr. Ox is not interested in lighting or in hydrogen. The project is a ruse that enables him to integrate himself in the town and perform his treacherous work in an undisturbed manner. His actual goal is to produce large amounts of oxygen in order to conduct a socio-psychological experiment on the town's population "in the interest of science." According to his hypothesis, the increased oxygen concentration in the air would change the mentality of rural people, by stirring up eagerness and aggressiveness, which he considers an improvement, an essential step to "reform the world." However, before Dr. Ox can start the experiment, the gas plant explodes by accident.

One might be tempted to read *Dr. Ox's Experiment* as a parable of contemporary industrialization: modern technology transforms not only the means of production but also the personality of the people. However, Dr. Ox is not an industrialist but a scientist who conducts a social experiment, or better, a chemical experiment on society under the guise of technological improvement. The change of mentality in the rural town is not a side effect but the actual goal of the project: common virtues, such as peacefulness, contentedness, and happiness, should turn into the vices of eagerness and aggressiveness. Because

Dr. Ox confuses vices and virtues in the name of science to reform the world, and because he cheats others in order to achieve his goals, he represents the morally perverted "mad scientist" whom no one should trust.

CRIMINAL INTENT AND POWER HUNGER

Death by poisoning (that is, using chemicals) has always been a popular theme in dramas, from Euripides' *Medea* (431 BC) and Shakespeare's *Hamlet* (1609) to modern detective novels. In most of these cases, the murderer's knowledge of chemistry is minimal or unnecessary. The "criminal chemist," on the other hand, systematically uses his expert knowledge of chemistry to commit crimes, not in the name of science, but for his personal enrichment or empowerment. The character can be traced back to the medieval tempter who first makes his victims addicted to alchemy by some chemical trickery and then sucks them dry financially. Whereas the late eighteenth century featured various comical versions of this trope, for instance in Giacomo Casanova's *The History of My Life* (1770–1772), Rudolf Erich Raspe's *Baron Munchausen* (1785), and the pseudonymously published *Confessions du Comte Cagliostro* (1787), nineteenth-century authors would employ this theme in the new genre of crime and detective novels.

For example, Wilkie Collins, generally considered the founder of British detective novels, wrote not only fully fleshed versions of the standard "mad scientist" (see above) but also the first modern version of the "criminal chemist." In his novel *The Haunted Hotel, a Mystery of Modern Venice* (1878), a talented young chemist, Baron Rivar, gives up his promising research career for the chemical pursuit of gold-making, which consumes both his own fortune and that of his sister. To cover the expenses of his experiments, he convinces her to marry a man whom they murder together to perpetrate a life insurance fraud. To that end, they show to the insurance company the dead body of the servant who has coincidentally died from bronchitis, while Baron Rivar uses his chemical skills to completely dissolve the corpse of the murdered man in acid.

Chemistry began to dominate the "criminal scientist" genre probably for two reasons. First, the crime theme was already existent in "mad alchemist" stories where the obsession with gold-making let the "mad alchemist" pursue his work by any means, including criminality. Unlike later authors, Collins simply reworked that old plot in *The Haunted Hotel*. Second, authors thought that chemistry had extraordinary capacities, at least in the view of their readers, which could make for interesting crime stories.

The most famous example of the second reason is H.G. Wells' novel *The Invisible Man* (1897); its numerous film and television series adaptions in the twentieth century speak for the great popularity of the theme. The main character, Griffin, who had once studied medicine and "won the medal for chemistry," moves to a small town and outfits a chemical laboratory to create a substance that would temporarily change the "refractive index" of the human body to that of air. He succeeds and becomes invisible upon taking the drug, which allows him to act incognito. Griffin's invisibility enables him to threaten and punish people anonymously, giving him a thirst for power, and a desire to establish a "reign of terror" in which he can impose any rule that pleases him.

While other nineteenth-century "criminal chemists" typically pursue riches, corresponding to the old "mad alchemist" theme, Griffin strives for power and domination. Already in 1871, the American author William Henry Rhodes, writing under the pseudonym "Caxton," had in his short story "The Case of Summerfield" (1871) a chemist who seeks power by threating to destroy the oceans with his new invention that enables him to "ignite" water. Both Caxton and Wells created the prototype for a popular character in science fiction during the twentieth century: the super-villain "mad scientist" who strives for world domination (Schummer 2022).

ABSENT-MINDED BENEVOLENCE

A few years after he had introduced the morally perverted chemist Dr. Ox, Jules Verne published another story that contained a prototype of the "benevolent and absent-minded scientist," who would in the mid-twentieth century become a popular character trope in adventure movies and comics series for children (Schummer 2022). In Verne's space adventure novel *Hector Servadac* (1877, *Off on a Comet*), a comet strikes the Earth and carries off a piece of the Earth with its inhabitants, including the hero Hector. When the group begins exploring the comet, they meet a man who once was Hector's science teacher. We first learn about Professor Rosette by Hector's memories of several funny stories in which Rosette was tricked by his pupils. For instance, while preparing a chemical experiment, Rosette was so preoccupied that Hector and his classmates could easily make fun of him by replacing some chemicals without Rosette noticing, so that the experiment did not work.

In the meantime Rosette had turned into an ambitious astronomer, who actually planned his travel on the comet by calculating the time and place of the strike. Although he is a strange person to deal with and at times absent-minded and eccentric, he turns out to be extremely helpful to the others. Thanks to his scientific knowledge, they learn about their location, the size and properties of the comet, and how to live on it. He calculates the comet's orbit and the next time when it comes close to Earth, which eventually helps them to return home.

The "benevolent and absent-minded scientist" still resembles the "mad scientist," with such characteristics as his absent-mindedness or obsession with his work, his eccentric and unsociable character, and at times a sinister appearance. However, when his interests harmonize with those of others, he is helpful, with surprising and wondrous capacities. He is never the main character, neither hero nor villain, but an essential actor in the plot, like a *deus ex machina* who allows authors to make sudden narrative turns in their stories. However, the "benevolent and absent-minded scientist" is rarely a chemist, but rather a general scientist-engineer, as if the character would not easily fit the stereotype of a chemist.

HEROIC CRIME INVESTIGATION

The negative image of fictional chemists throughout the nineteenth century was partly compensated for by a single character of extraordinary popularity, the private detective Sherlock Holmes. The British author and physician Arthur Conan Doyle introduced Holmes in his 1887 novel *A Study in Scarlet*, to be followed by three other novels and fifty-six short stories published during the next forty years. Right at the beginning of the first novel, Holmes is introduced as a passionate and ingenious experimental chemist who has just discovered a chemical reagent to detect traces of blood. The discovery serves his other great passion, resolving intricate criminal cases, by identifying blood stains on the clothes of possible murderers.

Much more than a mere instrument in crime investigation, chemistry prominently features in many stories involving Holmes. On the one hand, Holmes occasionally secludes himself overnight in his private laboratory to obsessively work on some chemical analysis, which his friend and assistant, Dr. Watson, always notices with bewilderment. On the other hand, chemical reasoning, as for instance in classical organic structure elucidation, as Holmes confesses many times, is the model on which he resolves his criminal cases, to the astonishment of fictional policemen and actual readers. By smartly interacting with the subjects and diligently observing even the smallest details, he eliminates all possible explanations until only one remains. There is probably no better approach to illustrate that chemical method to a broader audience than by applying it to exciting cases of criminal investigation.

Holmes is even more benevolent than Professor Rosette, and he is the main hero of all stories. And yet, Holmes is still a derivative of the "mad scientist." He is eccentric, obsessive in his work, cold-blooded in all other matters, not very sociable, and largely remains mysterious to others, including Watson, the narrator of most of the stories. However, the entire collection of stories also serves as an invitation and guide for understanding that type of person better, to acknowledge his eccentricity as a different but useful way of reasoning, rather than condemning it as a form of madness.

While Doyle himself created in 1912 Professor Challenger, an irascible biologist who experiences various adventures and mysterious events, other authors developed characters modeled on Holmes. Perhaps the most famous variation was chemistry professor Craig Kennedy, who appeared in numerous short stories and novels published between 1910 and 1918 by US author Arthur Benjamin Reeve. Nearly all the characteristic traits of the "mad scientist" have disappeared in Craig Kennedy, who embodied Reeve's vision of "criminal science" that would simply "apply science to the detection of crime, the same sort of methods by which you trace out the presence of a chemical" (*The Silent Bullet* 1910: 3).

ENVIRONMENTAL POLLUTION IN NOVELS

By the second half of the nineteenth century, European countries were well aware of the problems of pollution from chemical and other industries (Chapter 5; Homburg et al. 1998: 121–201). Although the issue of pollution would naturally have lent itself to the "mad scientist" trope by featuring chemists or chemical firms who in their obsession for some material perfection pollute the environment with a lack of circumspection, it is almost absent in the literature of the period. There were complaints about noise, smoke, smell, dust, lighting, and other human-made annoyances in metropolitan areas, and there was a new appreciation of untouched nature, from early-nineteenth-century German Romanticism to Henry David Thoreau's *Walden* (1854), which has recently been called "nature writing." However, the notion that nature as such could be severely affected, damaged, or endangered by industrialization was largely unknown in the nineteenth century.

One of the very rare early novels to focus on industrial pollution of the natural environment is *Pfisters Mühle* (1884, *Pfister's Mill*) by the German author Wilhelm Raabe. It does not deal with untouched nature, but with a recreational area close to Berlin, literally the natural environment of a big city. By a creek there is an old mill to which a restaurant catering to daily visitors from Berlin was once attached. Dr. Eberhard Pfister, a language teacher, inherited the mill from his father and sold it to a company. The firm has recently built a sugar plant upstream and wants to turn the mill into another plant. Pfister, who spent his youth there, visits the mill site to show it to his young wife. However, nothing is as it was before. The sugar industry has turned the creek into a sewer with dark brown foul-smelling water. The vegetation and fauna has completely changed, and the mill and restaurant are run down and abandoned.

Pfister asks his former mentor Dr. Asche (the name is a colloquial German term for money) for advice. Dr. Asche recently shifted his interests from ancient languages to modern chemistry. He does some chemical analysis but cannot help further. Instead, he realizes the financial opportunity to establish a huge factory for chemical dry cleaning in the surroundings of Berlin. Raabe here alluded to what was the biggest dry cleaning plant in Germany in 1873, W. Spindler, a few kilometers upstream from Berlin, whose business caused massive pollution.

Raabe's novel describes the process of industrialization using several parallel themes: from traditional craft to modern industry, from undisturbed beautiful nature to a malodorous sewer, from philology of ancient languages to the modern science of chemistry, and from the pursuit of wisdom to the longing for money. However, the novel was not well received and nearly fell into oblivion before it received renewed attention in the late twentieth century. The fictional treatment of environmental destruction by chemistry in *Pfister's Mill* remained a rarity in the nineteenth century, and would only re-emerge in the twentieth century.

Novelists continued to disregard the environmental theme well into the twentieth century (Schummer 2022). When it appeared in literary works after World War I, it was only in apocalyptic visions of total world destruction, first by poison gas and then by other scientific inventions such as nuclear energy and runaway bugs. Environmental pollution in the proper sense would only appear in the final decades of the twentieth century in the eco-thriller genre after the division of hero and villain had been sorted out anew, such that a scientist-hero fights the polluting chemical industry that has now assumed in part the role of the "mad scientist."

THE VISUAL REPRESENTATION OF CHEMISTS AND CHEMISTRY

Like the environmental pollution theme, the "mad scientist" theme is almost absent in nineteenth-century visual representation of chemistry, but for largely different reasons. First, the visual representations of chemistry in that period were mostly commissioned by chemists or chemical firms, as in portraits, technical drawings, and paintings of industrial landscapes. Second, just as the visual representation of the medieval "mad alchemist" appeared only later in illustrations of printed literary works, so did the nineteenth-century literary creations of "mad scientists" visually appear only in movie adaptions of the next century. Thus, the visual image lagged behind the literary image. However, once the motion picture industry was established, "mad scientist" movies began to dominate the visual image of science, starting with *Frankenstein* (1908) and *Dr Jekyll and Mr Hyde* (1910).

Considering that very little historical research has been done on visualization of science in the nineteenth century, the following can only briefly survey the visual representation of chemical laboratories, portraits, and industrial landscapes.

Following the French *Encyclopédie*, detailed technical drawings of laboratory apparatus and equipment became popular illustrations and important sources of information in encyclopedias, handbooks, and textbooks, commissioned by scientists and tailored to scientists, students, and interested laypersons. Nonetheless, independent artistic representations of laboratories, painted for a general audience, continued to depict medieval alchemical laboratories, from Joseph Wright of Derby's *The Alchemist Discovering Phosphorus* (1771/1795) to Carl Spitzweg's *The Alchemist* (*ca.* 1860). Thus, the iconography that originated from illustrations of the "mad alchemist" literature and which had been elaborated in the seventeenth century (see above) seems to have further dominated the public visual imagination of a chemical laboratory.

Because portraits of chemists were mostly commissioned by the depicted chemists themselves, or posthumously by their pupils and followers, they illustrate how chemists wanted to be seen at the time. In his pioneering work, Marco Beretta (2001) has pointed out the important role of portraiture – in paintings and engravings, on medals, and as statues – on shaping the historiography of chemistry. Starting with the famous painting *Portrait of Antoine-Laurent and Marie-Anne Lavoisier* (1788) commissioned by Marie-Anne Paulze Lavoisier and created by her art teacher Jacques-Louis David, portraits of Lavoisier at first served to promote his "chemical revolution." At the turn of the nineteenth century, portraits celebrated Lavoisier as a martyr of science, because of his arrest and execution by the Jacobins, which was actually for his privileged role as tax collector rather than as scientist. Later they contributed significantly to the creation of heroic legends about Lavoisier. As Beretta (2001: 57) put it: "In the iconographic representations, the tendency to idealize Lavoisier and his contribution and to see him as a legend grew throughout the whole nineteenth century, finally culminating in the monumental bronze statue in front of the *Eglise de la Madeleine* in 1900." Thus, contrary to the dominant public image of the "mad scientist" trope, not least in France, chemists tried to establish the legend of a heroic chemist.

Self-commissioned portraits of nineteenth-century chemists largely fall into four groups, depending on their depicted surroundings.[4] In the first group, chemists are shown sitting on a chair with some glassware or chemical apparatus in the background and books or notes in the foreground, echoing David's portrait of Lavoisier and indicating an emphasis on both experimentation and theorizing. In the second group, books or notes are the only props, pointing to their theoretical inclination. A third type of portrait shows the chemist at work in his laboratory, emphasizing the role of experimentation. In the fourth type of images, the chemist is depicted with, or proudly presents, his most important invention, reflecting their desire for technological improvement. For example, William Henry Perkin holds a skein dyed with the first synthetic aniline dye in his famous portrait of 1892. These four types of portraits illustrate both different personal orientations and different styles or scientific identities between the experimental, theoretical, and applied field.

The third type of portrait, the chemist at work in his laboratory, is particularly instructive about the dynamics of visual culture. The most common pose of that type is a chemist holding a flask at eye level and gazing at it, which would become the stereotypical image of a chemist in the twentieth century. This particular pose has a long iconographic history that shaped its cultural meaning (Schummer and Spector 2007). From the thirteenth to the sixteenth centuries, it was commonly found in portraits of doctors examining a urine flask. Originally it had signified uroscopy and had been an emblem of medicine. When uroscopy fell into disrepute, the pose became a symbol of quack medicine, then of fraudulent alchemy, and eventually of any kind of fraud. Artists widely used that pose in satire from the early sixteenth to the late nineteenth centuries, including satirical drawings of chemists. Chemists were most likely ignorant of the cultural history of that pose when they embraced an established symbol of fraud as a professional icon at the end of the nineteenth century.

Like portraits, images of chemical factories were usually commissioned. However, there are examples of non-commissioned nineteenth-century paintings of industrial landscapes, mostly of the iron industry, that provide some insight into independent artistic renditions (Schummer and Spector 2007). While the environmental effects of industry in the form of noise, smoke, and odors had already been noticed and publicly debated, artists tended to depict industry in one of two ways. Either they emphasized the smoke and the fire by colorful and mysterious renderings, reminiscent of a spectacular sunset and in the style of sublime landscape painting. Or they set the factory into a benign picturesque landscape. In the last type of images, regardless of whether they were commissioned or not, the smoking chimney was a common symbol of economic growth and prosperity, because for most people heating their home with an open fire that inevitably produced smoke was a luxury.

That artistic tradition was taken to an extreme in one of the first chemical plant paintings, Robert Friedrich Stieler's commissioned image of the BASF factory in Ludwigshafen, Germany from 1881.[5] The panoramic painting is filled with factory buildings and even suggests that they would continue in both directions off the frame, like a truncated view of a medium-sized city. It includes twenty-two smokestacks, most of them emitting black smoke, and about thirty sources of white smoke or steam, whereas the embedding landscape with a cloudy sky has receded into the background. Numerous workers and carriages on the streets, loading and unloading boats on the riverfront, and an incoming freight train emphasize vibrant economic activity. What might strike us today as a source of tremendous environmental pollution was in the late nineteenth century still an artistic symbol of enormous prosperity and growth.

CONCLUSION

The Literary Image of Chemistry

Before the rise of visual media such as movies and television, literature clearly dominated the public image of science. Nineteenth-century visual representations of chemistry, which were mostly commissioned works and lacked any reference to the "mad scientist" trope, hardly reached out to a broader public. If they were meant to correct the literary image in the form of heroic portraiture, they had little to no effect. Indeed, chemists' attempt to create a visual image of themselves was so disconnected from the general discourse that they (accidentally) embraced an age-old symbol of fraud as their professional icon.

During the nineteenth century, writers shaped a distinctive multifaceted literary image of chemists and chemistry that would have a long-lasting impact. They did so not only by "mad scientist" stories proper, but also by including characters in their narratives who are only said to be chemists and by letting other characters express their views about chemistry. By putting the "mad scientist" in this broader context of the overall image of chemistry in literature we can, in conclusion, try to understand why chemistry was considered so special among all the disciplines.

As the "mad scientist" stories suggest, chemistry is acknowledged as having great power and could bring about wondrous results that go beyond anything imagined before. For instance, the Count in Wilkie Collins' *The Woman in White* (1860) argues, "Chemists – I assert it emphatically – might sway, if they pleased, the destinies of humanity. ... Mind, they say, rules the world. But what rules the mind? The body (follow me closely here) lies at the mercy of the most omnipotent of all potentates – the Chemist."

Chemists are also described as particularly intelligent in their scientific reasoning, which is proved not only by their results but also in the way they argue, conclude, and explain. However, their passion for science is typically contrasted with a lack of interest in literature, the arts, and humanities, as well as with a certain clumsiness in human affairs and with social seclusion. That combination makes them suspect in many regards and subject to various associations that threaten prevailing metaphysical and religious views.

In a modest form, chemistry is associated with skepticism and nihilism. For instance, the German writer Karl Ferdinand Gutzkow had his character Oleander in *Die Ritter vom Geiste* (1850/1851, *Knight of the Mind*) argue against the school of critical philosophy: " ... these philosophers of the absolute Nothing are the Liebigs of the invisible world .... Such as the chemical retort invents element after element, each being decomposed over and again, such does the philosophical, heartless intellect of the school resolve Everything into the perfect Nothing by criticism ... even believing that the immortality of the soul would have been disproved." It is therefore no coincidence that the arch-nihilist Bazarov in Ivan Turgenev's *Fathers and Sons* (1862) is a passionate chemist who takes the received metaphysical ideas apart like the chemist doing an elemental analysis of a substance. His skepticism is paired with contempt for the traditional wisdom: "A decent chemist is twenty times more useful than any poet."

Chemistry is also sometimes associated with the protagonists of the Enlightenment movement and the French Revolution. The association goes via the Order of Illuminati (Latin for "the enlightened"), a secret society founded by Adam Weishaupt at the University of Ingolstadt in Bavaria in the late 1770s, about which there were strong rumors in the nineteenth century that they had instigated the French Revolution. For instance, in Alexandre Dumas' *Joseph Balsamo* (1844–1846) already mentioned, the "mad scientist" Althotas is said to be the head of the Illuminati, aiming at overthrowing the monarchs of Europe. And Mary Shelley placed Frankenstein at the University of Ingolstadt certainly for no other reason than to establish that link.

Chemists are frequently portrayed as being both materialists and atheists, i.e. they do not believe in the existence of a soul or God, which for many nineteenth-century authors were views closely connected to each other. Unlike other sciences, chemistry is described to convey such a worldview. For instance, after a brief dialogue between a chemist and a physicist in *La Peau de Chagrin* (1831, *The Magic Skin*), Balzac explains: "The universe for a mechanician is a machine that requires an operator; for chemistry – that fiendish employment of decomposing all things – the world is a gas endowed with the power of movement." In Dostoyevsky's *Brothers Karamazov* (1879) Mitya Karamazov learns about neurochemistry and then desperately confesses to his brother and priest: "I am sorry to lose God ... [and the belief that] I've got a soul, and that I am some sort of image and likeness. ... It's chemistry, brother, chemistry! There's no help for it, your reverence, you must make way for chemistry."

Many Christian authors combined atheism or agnosticism with the hubris theme by having chemists daring to "play God," which in the Christian mythology of the Fallen Angels was the original crime of Satan. Indeed, this appears to be an overarching theme of all "mad scientist" stories, reminiscent of a prominent accusation of alchemy in the Middle Ages: the chemist who aspires to create material improvements in his laboratory is "playing God." For instance, in Balzac's already mentioned *La Recherche de l'absolu* (1834), Claes argues with his wife:

> "I shall make metals," he cried; "I shall make diamonds, I shall be a co-worker with Nature!" "Will you be the happier?" she asked in despair. "Accursed science! Accursed demon! You forget, Claes, that you commit the sin of pride, the sin of which Satan was guilty; you assume the attributes of God."

Beyond hubris comes Satanism and megalomania. In *La Peau de Chagrin*, Balzac let his chemist, Baron Japhet, straightforwardly confess: "I believe in the devil." And Alexandre Dumas has his Dr. Sturler, believing that he has found a poison's antidote, pronounce: "Am I not God like God – more God than God since I can retake and give back life, cause death to be born, and destroy death?" (*Le Comte Hermann*, 1849, Epilogue).

In general, the portrayal of chemists in nineteenth-century literature contrasts their scientific intelligence and power over nature with their naiveté and corresponding deficit in moral reasoning. The most common character trait, indeed part of the standard form of the "mad scientist," is the lack of circumspection and precaution, based on the misunderstanding that good intentions alone are important in moral matters. This literary image of chemists became firmly established during the first half of the nineteenth century, such that an author would only need to mention in passing that a character is a chemist, or has a chemical laboratory, to connect to the trope.

Why Chemistry?

Why did so many nineteenth-century authors, including the most famous and influential ones, develop such a negative image of chemistry among all the sciences? Today's readers might be inclined to see these negative portrayals as early warnings of environmental problems and disasters caused by the chemistry industry. However, the "mad scientist" was developed in the early nineteenth century, when the chemical industry hardly existed. When it actually existed and environmental issues were publicly debated later in the century, writers showed almost no interest in those topics, as we have seen above. Moreover, it is difficult to find a single historical example of a real scientist who would meet the criteria of the "mad scientist" as portrayed in the literature. Therefore, actual experience could not have served as a model. Nor can we celebrate Mary Shelley and other authors as ingenious prophets of a future industrial age that was totally beyond their imagination. Our answer to the question must therefore be more complex, and consider the general intellectual context of the time.

One reason for the emphasis on chemistry is the long-standing literary heritage of the "mad alchemist," a well-developed and handy character of satire that could easily be revived and adjusted to contemporary science. The fact that "mad scientists" are still today frequently surrounded by alchemical apparatus supports this thesis, but why was the "mad alchemist" revived at the turn to the nineteenth century and specifically directed at chemistry?

One answer might be found in the shifting ground of scientific disciplines during the nineteenth century, when the traditional religious-based educational system at universities was transformed into professional training in the different branches of science. In the new university system, both chemistry and physiology established laboratory experimentation as a new style of scientific research that no longer depended on received metaphysical or religious foundations. Because chemists and physicians performing chemical experiments are the main characters of nineteenth-century "mad scientist" stories, it is likely that many authors felt uneasy with this kind of science that tried to be "independent" of metaphysics. Thus, they recurrently described their fictional chemists as being materialist, nihilist, and atheist, although contemporary chemists provided little historical evidence for that.[6] Rather than the individual researcher, it was the experimental approach that exclusively focused on material events (thus, "materialism") and intentionally excluded metaphysical and religious ideas in explanations (thus, "nihilism," "atheism"), which made chemistry and physiology suspect. Moreover, because nineteenth-century authors used the same attributes (materialism, nihilism, and atheism) to discredit the Enlightenment, they could metaphorically relate that movement to chemistry, even though prominent chemists like Lavoisier actually represented the old political system.

Much of nineteenth-century literature was written for the moral education of adults, and many works express extremely conservative, if not fundamentalist, Christian ideas. The very notion that science could change nature in order to improve the material conditions of life, by inventing some new useful materials or by developing a medicine against an illness, raised the age-old accusation of "playing God," with runs through most of the "mad scientist" stories. Moreover, because for many Christians the Bible remained the only basis of morality, a science or medicine that was not based on religion could easily result in evil or crime. That seems to be the main reason for all the variations of moral failure by "mad scientists."

Furthermore, for much of the nineteenth century, poets, novelists, and playwrights represented the humanities or even wisdom, at least in public life, more than philosophers, historians, and philologists. The rise of the intellectual and economic status of the natural sciences, particularly chemistry with its rapid discovery of new elements and apparent powers of creation, undoubtedly caused a certain rivalry and envy, and posed a threat to the traditional status of the humanities as the arbiter of morality.

Finally, once the "mad scientist" was established, it took a literary life of its own. Compared to previous centuries, late-nineteenth-century writers were extremely productive in inventing new literary styles and genres. Establishing a new style or genre works best with classical or well-known themes that allow readers to more easily recognize the novel from the traditional. To that end, the "mad scientist" trope proved extremely successful in various literary genres, from romance to crime to science fiction.

NOTES

CHAPTER 5

1. Paxton had designed glasshouses for the Duke of Devonshire at Chatsworth House (Yorkshire, England). The exhibition building was given the name "Crystal Palace" by Mark Lemon, editor of *Punch*, because of its crystal appearance in sunlight.
2. Charles Goodyear had discovered vulcanization of rubber in America in 1839.
3. It was in Manchester in 1848 that R.S. Edelston caught the first live specimen of the dark *cabonaria* form of the peppered moth, a variety of the species that could survive predators when resting on soot-covered surfaces.

CHAPTER 7

1. Other early institutes include Ipswich (1824), Manchester (1824), Cheltenham (1834), and Gloucester (1840).
2. In 1904, Rayleigh was awarded the Nobel Prize in physics and Ramsay the Nobel Prize in chemistry for their work on the densities of gases and their discovery of argon.
3. Committees were often called after their chairman: the Samuelson Committee after its chairman Bernard Samuelson.
4. Harvard College was a separate institution and by 1858 had a laboratory for teaching practical chemistry and experienced teaching staff.
5. The term "fume hood" is still used in the USA.
6. In 1914 the journal split into *Annales de Chimie* and *Annales de Physique*.

CHAPTER 8

1. Works that go beyond allegorical alchemy include Read (1947), Krätz (1990), Haynes (1994), Linden (1996), Meakin (1995), Schummer (2006), Schummer et al. (2007), Schummer (2008), Labinger (2011), and Ziolkowski (2015).
2. For the images discussed in this section, see Schummer and Spector (2007).

3   As Balzac explained in a letter: "Le héros de *La Recherche de l'absolu* représente tous les efforts de la chimie moderne" (the heroes of *La Recherche de l'absolu* represent all the efforts of modern chemistry) (Balzac in Ambrière 1999: 401).
4   For exemplary images and more details, see Schummer and Spector (2007).
5   Many later online reproductions of the paintings have cut off parts or rendered the image such that less smoke is visible.
6   For instance, in the first half of the nineteenth century most leading chemists believed in one or the other form of vital forces. Pasteur, one of the most prominent chemists of the second half, was deeply religious and devoted much time to disproving "spontaneous generation," which in his view would have undermined the belief in a creator God.

BIBLIOGRAPHY

Abney Salomon, Charlotte A. 2019. "The Pocket Laboratory: The Blowpipe in Eighteenth-Century Swedish Chemistry." *Ambix*, 66: 1–22.
Accum, Frederick. 1819. *Description of the Process of Manufacturing Coal-Gas*. London: Thomas Boys.
Accum, Frederick. 1824. *Dictionary of Chemical and Philosophical Apparatus*. London: Thomas Boys.
Agar, Jon. 2012. *Science in the Twentieth Century and Beyond*. Malden, MA: Polity.
Ambrière, Madeleine. 1999. *Balzac et la Recherche de l'absolu*. Paris: Presses Universitaires de France.
American Chemical Society. 1997. *Production of Aluminum: The Hall–Héroult Process*. Washington, DC: American Chemical Society.
Angulo, A. 2012. "The Polytechnic Comes to America: How French Approaches to Science Instruction Influenced Mid-Nineteenth Century American Higher Education." *Annals of Science*, 50: 315–38.
Archer, Mary D., and Haley, Christopher D. 2005. *The 1702 Chair in Chemistry at Cambridge*. Cambridge: Cambridge University Press.
Argles, Michael. 1964. *South Kensington to Robbins. An Account of English Technical and Scientific Education Since 1851*. London: Longmans.
Averley, Gwen. 1986. "The 'Social Chemists': English Chemical Societies in the Eighteenth and Early Nineteenth Century." *Ambix*, 33: 99–128.
Ball, Philip. 2003. *Bright Earth: Art and the Invention of Color*. Chicago, IL: University of Chicago Press.
Ballard, Dr. Edward. 1878. "Supplement Containing the Report of the Medical Officer for 1876." *Sixth Annual Report of the Local Government 1876–77*. London: Her Majesty's Stationery Office.
Barker, T.C., R. Dickinson, and D.W.F. Hardie. 1956. "The Origins of the Synthetic Alkali Industry in Britain." *Economica*, 23: 158–71.
Barker, Theodore C. 1977. *The Glassmakers. Pilkington: The Rise of an International Company, 1826–1976*. London: Weidenfeld and Nicolson.
Beardsley, Edward H. 1964. *The Rise of the American Chemistry Profession, 1850–1900*. Gainesville, FL: University of Florida Press.

Becker, Barbara J. 2011. *Unravelling Starlight: William and Margaret Huggins and the Rise of the New Astronomy*. Cambridge: Cambridge University Press.

Beer, John Joseph. 1959. *The Emergence of the German Dye Industry*. Champaign, IL: University of Illinois Press.

Ben-David, Joseph. 1971. *The Scientists' Role in Society: A Comparative Study*. Englewood Cliffs, NJ: Prentice-Hall.

Benfey, Otto T. (ed.). 1963. *Classics in the Theory of Chemical Combination*. New York: Dover Publications.

Bennet, Abraham. 1787. "Description of a New Electrometer. In a Letter From the Rev. Abraham Bennet, M. A. To the Rev. Joseph Priestley, LL.S. F. R. S." *Philosophical Transactions of the Royal Society*, 77: 26–31.

Bensaude-Vincent, Bernadette. 1986. "Mendeleev's Periodic System of Chemical Elements." *British Journal for the History of Science*, 19: 3–17.

Bensaude-Vincent, Bernadette. 1990. "A View of the Chemical Revolution Through Contemporary Textbooks: Lavoisier, Fourcroy and Chaptal." *British Journal for the History of Science*, 23: 435–60.

Bentley, Jonathan. 1970. "The Chemical Department of the Royal School of Mines. Its Origins and Development Under A.W. Hofmann." *Ambix* 17: 153–81.

Beretta, Marco. 2001. *Imaging a Career in Science: The Iconography of Antoine Laurent Lavoisier*. Canton, MA: Science History Publications.

Beretta, Marco. 2014. "Between the Workshop and the Laboratory: Lavoisier's Network of Instrument Makers." *Osiris*, 29: 197–214.

Bertomeu-Sánchez, José Ramón. 2015. "Chemistry, Microscopy and Smell: Bloodstains and Nineteenth-Century Legal Medicine." *Annals of Science*, 72: 490–516.

Bertomeu-Sánchez, José Ramón, and Agustí Nieto-Galan (eds). 2006. *Chemistry, Medicine, and Crime: Mateu J.B. Orfila (1787–1853) and His Times*. Sagamore Beach, MA: Science History Publications.

Bertrams, Kenneth, Ernst Homburg, and Nicolas Coupain. 2013. *Solvay: The History of a Multinational Family Firm*. New York: Cambridge University Press.

Berzelius, Jakob. 1814. "Experiments to Determine the Definite Proportions in Which the Elements of Organic Nature Are Combined." *Annals of Philosophy*, 4: 323–31, 401–409.

Berzelius, Jakob. 1841. *Sieben Kupfertafeln vom zehnten Bande von Berzelius' Lehrbuch der Chemie: chemische Operationen und Geräthschaften*. Dresden and Leipzig: Arnoldische Buchhandlung.

Bigelow, W.D. 1898. "The Development of Pure Food Legislation." *Science*, 7/172: 505–13.

Blakemore, Harold. 1974. *British Nitrates and Chilean Politics, 1886–1896*. London: Athlone Press.

Blanc, Paul David. 2016. *Fake Silk: The Lethal History of Viscose Rayon*. New Haven, CT: Yale University Press.

Bogert, Marston Taylor. 1931. "Charles F. Chandler." *Biographical Memoirs of the National Academy of Sciences*.

Borscheid, Peter. 1976. *Naturwissenschaft, Staat und Industrie in Baden (1848–1914)*. Stuttgart: Klett.

Bradley, Margaret. 1976. "The Facilities for Practical Instruction in Science During the Early Years of the École Polytechnique." *Annals of Science*, 33: 425–46.

Brock, William H. 1967. "The London Chemical Society 1824." *Ambix*, 14: 133–9.

Brock, William H. 1969. "Studies in the History of Prout's Hypothesis." *Annals of Science*, 25: 49–80, 127–137.
Brock, William H. 1985. *From Protyle to Proton: William Prout and the Nature of Matter, 1785–1985*. Bristol: Adam Hilger.
Brock, William H. 1993. *The Norton History of Chemistry*. New York: Norton.
Brock, William H. 1997. *Justus Liebig: The Chemical Gatekeeper*. Cambridge: Cambridge University Press.
Brock, William H. 2008. *William Crookes (1832–1919) and the Commercialization of Science*. Burlington, VT: Ashgate.
Brooke, John Hedley. 1968. "Wöhler's Urea, and its Vital Force: A Verdict from the Chemists." *Ambix*, 15: 84–114.
Brooke, John Hedley. 1973. "Chlorine Substitution and the Future of Organic Chemistry: Methodological Issues in the Laurent–Berzelius Correspondance (1843–1844)." *Studies in History and Philosophy of Science*, 4: 47.
Brooke, John Hedley. 1981. "Avogadro's Hypothesis and its Fate: A Case Study in the Failure of Case Studies." *History of Science*, 19: 235–73.
Brooke, John Hedley. 1991. *Science and Religion: Some Historical Perspectives*. Cambridge: Cambridge University Press.
Brooke, John Hedley. 2007. "Overtaking Nature? The Changing Scope of Organic Chemistry in the Nineteenth Century." In Bernadette Bensaude-Vincent and William R. Newman (eds), *The Artificial and the Natural: An Evolving Polarity*. Cambridge, MA: MIT Press.
Brooke, John Hedley, and Geoffrey N. Cantor. 2000. *Reconstructing Nature: The Engagement of Science and Religion*. New York: Oxford University Press.
Brooks, Nathan M. 1995. "Russian Chemistry in the 1850s: A Failed Attempt at Institutionalization." *Annals of Science*, 52: 577–89.
Brooks, Nathan M. 1998a. "Alexander Butlerov and the Professionalization of Science in Russia." *The Russian Review*, 57: 10–24.
Brooks, Nathan M. 1998b. "The Evolution of Chemistry in Russia During the Eighteenth and Nineteenth Centuries." In David Knight and Helge Kragh (eds), *The Making of the Chemist: The Social History of Chemistry in Europe, 1789–1914*. Cambridge: Cambridge University Press.
Brown, Alexander Crum. 1864. "On the Theory of Isomeric Compounds." *Transactions of the Royal Society of Edinburgh*, 23: 707–19.
Brush, Stephen G. 1996. "The Reception of Mendeleev's Periodic Law in America and Britain." *Isis*, 87: 595–628.
Bud, Robert, and Gerrylynn K. Roberts. 1984. *Science versus Practice. Chemistry in Victorian Britain*. Manchester: Manchester University Press.
Bud, Robert, and Deborah Jean Warner (eds). 1998. *Instruments of Science: An Historical Encyclopedia*. London/Washington, DC: The Science Museum and the Smithsonian Institution.
Burchfield, Joe D. 1975. *Lord Kelvin and the Age of the Earth*. Chicago, IL: University of Chicago Press.
Burns, D. Thorburn et al. 2014. *Important Figures in Analytical Chemistry from Germany in Brief Biographies*. New York: Springer.
Büttner, J. 2000. "Justus von Liebig and his Influence on Clinical Chemistry." *Ambix*, 47: 96–117.
Calkins, Alonzo. 1871. *Opium and the Opium Appetite*. Philadelphia, PA: Lippincott.
Campbell, W. Alec. 1971. *The Chemical Industry*. London: Longman.

Campos, Luis. 2007. "The Birth of Living Radium." *Representations*, 97: 1–27.
Campos, Luis. 2015. *Radium and the Secret of Life*. Chicago, IL: University of Chicago Press.
Cardwell, D.S.L. 1972. *The Organisation of Science in England*. London: Heinemann Educational Books.
Clayton, Edwyn Godwin. 1908. *Arthur Hill Hassall: Physician, and Sanitary Reformer*. London: Baillière, Tindall and Cox.
Clow, Archibald, and Nan L. Clow. 1952. *The Chemical Revolution. A Contribution to Social Technology*. London: The Batchworth Press.
Coffey, Patrick. 2008. *Cathedrals of Science: The Personalities and Rivalries that Made Modern Chemistry*. Oxford: Oxford University Press.
Coleman, William. 1971. *Biology in the Nineteenth Century: Problems of Form, Function, and Transformation*. Cambridge: Cambridge University Press.
Coley, Noel G. 1996. "Studies in the History of Animal Chemistry and its Relation to Physiology." *Ambix*, 43: 164–87.
Craig, Norman C. 1986. "Charles Martin Hall—The Young Man, his Mentor and his Metal." *Journal of Chemical Education*, 63: 557–9.
Crass, M.F. Jr. 1941a. "A History of Match Making: Part V—Safety or Strike-on-box Matches." *Journal of Chemical Education*, 18: 316–19.
Crass, M.F. Jr. 1941b. "A History of Match Making: Part IX—Phosphorus Necrosis." *Journal of Chemical Education*, 18: 428–31.
Creese, Mary R.S. 1991. "British Women of the Nineteenth and Early Twentieth Centuries who Contributed to Research in the Chemical Sciences." *British Journal for the History of Science*, 24: 275–305.
Crosland, Maurice P. 1978. *Gay-Lussac, Scientist and Bourgeois*. New York: Cambridge University Press.
Crosland, Maurice. 2003. "Difficult Beginnings in Experimental Science at Oxford: The Gothic Chemistry Laboratory." *Annals of Science*, 60: 399–421.
Cunningham, Andrew, and Nicholas Jardine (eds). 1990. *Romanticism in the Sciences*. Cambridge: Cambridge University Press.
Dana, Edward S. 1920. "Biographical Memoir of George Jarvis Brush (1831–1912)." *Memoir of National Academy of Sciences*, 17: 106–11.
Davy, John (ed.). 1840. *Bakerian Lectures and Miscellaneous Papers from 1806 to 1815*. London: Smith, Elder.
Davy, Humphry. 1968. "Presidential Address on the Occasion of the Presentation of the First Royal Medal of the Royal Society to John Dalton (1827)." In David M. Knight (ed.), *Classical Scientific Papers: Chemistry*. New York: American Elsevier Pub. Co.
DeBoer, George E. 1991. *A History of Ideas in Science Education. Implications for Practice*. New York: Teachers College Press.
Deltete, Robert J. 1999. "Helm and Boltzmann: Energetics At the Lübeck Naturforscherversammlung." *Synthese*, 119: 45–48.
Dingle, A.E. 1982. "'The Monster Nuisance of All': Landowners, Alkali Manufacturers, and Air Pollution, 1828–64." *Economic History Review*, 35: 529–48.
Dolan, Brian. 1998. "Blowpipes and Batteries: Humphry Davy, Edward Daniel Clarke, and Experimental Chemistry in Early Nineteenth-Century Britain." *Ambix*, 45: 137–62.
Dolan, Brian. 2003. "Embodied Skills and Travelling Savants: Experimental Chemistry in Eighteenth-Century Sweden and England." In A.A. Simões, A. Carneiro, and M.P. Diogo (eds), *Travels of Learning. A Geography of Science in Europe*. Dordrecht: Kluwer.

Dolby, R.G.A. 1976. "Debates Over the Theory of Solutions: A Study of Dissent in Physical Chemistry in the English-Speaking World in the Late Nineteenth and Early Twentieth Centuries." *Historical Studies in the Physical Sciences*, 7: 297–404.

Dolby, R.G.A. 1984. "Thermochemistry versus Thermodynamics: The Nineteenth Century Controversy." *History of Science*, 22: 375–400.

Dyer, Daniel, and Davis Gross. 2001. *The Generations of Corning: The Life and Times of a Global Corporation*. New York: Oxford University Press.

Eichengreen, Barry. 2003. "Monetary, Fiscal and Trade Policies in the Development of the Chemical Industry'." In Ashish Arora et al. (eds), *Chemicals and Long-Term Growth: Insights from the Chemical Industry*. New York: Wiley/Chemical Heritage Foundation.

Elliott, Paul. 1999. "Abraham Bennet, F.R.S. (1749–1799): A Provincial Electrician in Eighteenth-Century England." *Notes and Records of the Royal Society of London*, 53: 59–78.

Emmet, John P. 1826. "Letter to Thomas Jefferson, 27 May 1826."

Emsley, John. 2000. *The Shocking History of Phosphorus: A Biography of the Devil's Element*. London: Macmillan.

Emsley, John. 2003. *Nature's Building Blocks: An A–Z Guide to the Elements*. Oxford: Oxford University Press.

Fant, Kene. 1993. *Alfred Nobel: A Biography*. New York: Arcade.

Faraday, Michael. 1827. *Chemical Manipulation: Being Instructions to Students in Chemistry, on the Methods of Performing Experiments of Demonstration Or of Research, With Accuracy and Success*. London: W. Phillips.

Faraday, Michael. 1861. *The Chemical History of a Candle*. London: Chatto and Windus.

Farley, John, and Gerald Geison. 1974. "Science, Politics and Spontaneous Generation in Nineteenth Century France: The Pasteur–Pouchet Debate." *Bulletin of the History of Medicine*, 48: 161–98.

Flanders, Judith. 2003. *Inside the Victorian Home*. New York: Norton.

Fleming, James Rodger. 1998. *Historical Perspectives on Climate Change*. Oxford: Oxford University Press.

Flick, Carlos. 1980. "The Movement for Smoke Abatement in 19th-Century Britain." *Technology and Culture*, 21: 29–50.

Forster, Sam Vettese, and Robert M. Christie. 2013. "The Significance of the Introduction of Synthetic Dyes in the Mid-19th Century on the Democratisation of Western Fashion." *JAIC—Journal of the International Colour Association*, 11: 1–17.

Forster, Simon P., Anthony S. Travis, and Stefan Seeger. 2020. "Saccharin Beyond Serendipity: A German–American *Wechselspiel* of Invention and Industry." In S. Katzir et al. (eds), *Made in Germany: Technologie, Geschichte, Kultur*. Göttingen: Tel Aviv Yearbook for German History.

Fox, Nicholas J. 1988. "Scientific Theory Choice and Social Structure, the Case of Joseph Lister's Antisepsis, Humoral Theory and Asepsis." *History of Science*, 26: 367–97.

Fox, Robert. 2012. *The Savant and the State: Science and Cultural Politics in Nineteenth Century France*. Baltimore, MD: Johns Hopkins University Press.

Fox, Robert. 2016a. *Science Without Frontiers: Cosmopolitanism and National Interests in the World of Learning*. Corvallis: Oregon State University Press.

Fox, Robert. 2016b. "Science, Celebrity, Diplomacy: The Marcellin Berthelot Centenary, 1927." *Revue d'histoire des sciences et de leurs applications*, 69: 77–115.

Fox, Robert, and George Weisz (eds). 1980. *The Organization of Science and Technology in France 1808–1914*. Cambridge: Cambridge University Press.

Freemantle, Michael. 2002. "Cambridge Marks 300 Years of Chemistry." *Chemical and Engineering News*, 80: 39–43.

Freer, Paul C. 1895. *The Elements of Chemistry*. Boston, MA: Allyn and Bacon.

Friedel, Robert. 1983. *Pioneer Plastic: The Making and Selling of Celluloid*. Madison: University of Wisconsin Press.

Fruton, Joseph S. 1972. *Molecules and Life: Historical Essays on the Interplay of Chemistry and Biology*. New York: Wiley.

Fruton, Joseph S. 1985. "Contrasts in Scientific Style. Emil Fischer and Franz Hofmeister: Their Research Groups and their Theory of Protein Structure." *Proceedings of the American Philosophical Society*, 129: 313–70.

Fruton, Joseph S. 1988. "The Liebig Research Group: A Reappraisal." *Proceedings of the American Philosophical Society*, 132: 1–66.

Fruton, Joseph S. 1990. *Contrasts in Scientific Style: Research Groups in Chemical and Biological Sciences*. Philadelphia, PA: American Philosophical Society.

Gage, John. 1993. *Colour and Culture: Practice and Meaning From Antiquity to Abstraction*. London: Thames and Hudson.

Galloway, Robert. 1858. *A Manual of Qualitative Analysis*. London: J. Churchill.

Ganzenmüller, Wilhelm. 1938. *Die Alchemie im Mittelalter*. Paderborn: Bonifacius.

Garfield, Simon. 2000. *Mauve: How One Man Invented a Colour that Changed the World*. London: Faber & Faber.

Gay, Hannah. 2017. *The History of Imperial College London, 1907–2007: Higher Education and Research in Science, Technology and Medicine*. London: Imperial College Press.

Gay, Hannah, and Griffith, William P. 2017. *The Chemistry Department At Imperial College: A History, 1845–2000*. London: World Scientific Publishing.

Gay-Lussac, Joseph Louis, and Louis Jacques Thenard. 1811. *Recherches physico-chimiques,* Vol 1 Paris: Deterville.

Gee, B., and William H. Brock. 1991. "The Case of John Joseph Griffin: From Artisan-Chemist and Author-Instructor to Business Leader." *Ambix*, 38: 29–62.

Geison, Gerald. 1995. *The Private Science of Louis Pasteur*. Princeton, NJ: Princeton University Press.

Golinski, Jan. 1992. *Science as Public Culture: Chemistry and Enlightenment in Britain, 1760–1820*. Cambridge: Cambridge University Press.

Golinski, Jan. 1994. "Precision Instruments and the Demonstrative Order of Proof in Lavoisier's Chemistry." *Osiris*, 2nd Series: 30–47.

Gordin, Michael D. 2004. *A Well-Ordered Thing: Dmitrii Mendeleev and the Shadow of the Periodic Table*. New York: Basic Books.

Gordin, Michael D. 2008. "The Heidelberg Circle: German Inflections on the Professionalization of Russian Chemistry in the 1860s." *Osiris*, 23: 23–49.

Gordin, Michael D. 2015. *Scientific Babel: How Science Was Done Before and After Global English*. Chicago, IL: University of Chicago Press.

Gordin, Michael D. 2018. "Paper Tools and Periodic Tables: Newlands and Mendeleev Draw Grids." *Ambix*, 65: 30–51.

Graham, Loren R. 1993. *Science in Russia and the Soviet Union*. New York: Cambridge University Press.

Greene, John C. 1984. *American Science in the Age of Jefferson*. Ames: Iowa State University Press.

Gregory, Frederick. 1977. *Scientific Materialism in Nineteenth Century Germany*. Dordrecht: Reidel.
Griffel, Margaret Ross. 1990. *Operas in German: A Dictionary*. London: Greenwood Press.
Guralnick, Stanley M. 1979. "The Contexts of Faraday's Electrochemical Laws." *Isis*, 70: 59–75.
H. N., F. 1901. "Some Scientific Centres: The Laboratory of Wilhelm Ostwald." *Nature*, 64: 428.
Haber, Ludwig F. 1969. *The Chemical Industry During the Nineteenth Century: A Study of the Economic Aspect of Applied Chemistry in Europe and North America*. Oxford: Clarendon Press.
Haines, George, IV. 1958. "German Influence upon Scientific Instruction in England, 1867–1887." *Victorian Studies*, 1: 215–44.
Hamlin, Christopher. 1990. *A Science of Impurity: Water Analysis in Nineteenth Century Britain*. Berkeley, CA: University of California Press.
Hammond, P. W., and Harold Egan. 1992. *Weighed in the Balance*. London: HMSO.
Hannaway, Owen. 1976. "The German Model of Chemical Education in America: Ira Remsen at Johns Hopkins (1876–1913)." *Ambix*, 23: 145–64.
Hardie, D.W.F., and J. Davidson Pratt. 1966. *A History of the Modern British Chemical Industry*. Oxford: Pergamon.
Harwood, Jonathan. 2005. *Technology's Dilemma: Agricultural Colleges Between Science and Practice in Germany, 1860–1934*. Oxford: P. Lang.
Haynes, Roslynn D. 1994. *From Faust to Strangelove: Representations of Scientists in Western Literature*. Baltimore, MD: Johns Hopkins University Press.
Haynes, William. 1959. *Brimstone, the Stone that Burns: The Story of the Frasch Sulphur Industry*. Princeton, NJ: Van Nostrand.
Hearnshaw, John. 2010. "Auguste Comte's Blunder: An Account of the First Century of Stellar Spectroscopy and how it Took One Hundred Years to Prove that Comte was Wrong!" *Journal of Astronomical History and Heritage*, 13: 90–104.
Hendersen, W.O. 2006. *Industrial Revolution on the Continent: Germany, France, Russia, 1800–1914*. New York: Routledge.
Henry, William. 1815. *The Elements of Experimental Chemistry*. London: Baldwin, Cradock and Joy.
Hentschel, Klaus. 2002. *Mapping the Spectrum: Techniques of Visual Representation in Research and Teaching*. New York: Oxford University Press.
Hepler-Smith, Evan. 2015. "'Just as the Structural Formula Does': Names, Diagrams, and the Structure of Organic Chemistry at the Geneva Nomenclature Congress." *Ambix*, 62: 1–28.
Hirota, Noburo. 2016. *A History of Modern Chemistry*. Kyoto: Kyoto University Press.
Hirsh, Richard F. 1981. "A Conflict of Principles: The Discovery of Argon and the Debate over its Existence." *Ambix*, 28: 121–30.
Hodgson, Barbara. 2001. *In the Arms of Morphius: The Tragic History of Laudanum, Morphine, and Patent Medicines*. Buffalo, NY: Firefly Books.
Hoechst. 1985. *Farb Werke. Historische Etiketten. Historic Labels. Etiquettes Historiques. Etiquetas Históricas*. Frankfurt am Main: Hoechst Aktien Gesellschaft.
Hoff, J.H. van 't. 1874. *Voorstel tot uitbreiding der tegenwoordig in de scheikunde gebruikte structuur-formules in de ruimte; benevens een daarmeê samenhangende opmerkung omtrent het verband tusschen optisch actief vermogen en chemische constitutie van organische verbindingen*. Utrecht: Greven.

Hofmann, August W. 1865. "On the Combining Power of Atoms." *Proceedings of the Royal Institution of Great Britain*, 4: 410–30.
Hofmann, August W. 1876. *The Faraday Lecture for 1875: The Life-Work of Liebig in Experimental and Philosophic Chemistry*. London: MacMillan.
Holmes, Frederic L. 1962. "From Elective Affinities to Chemical Equilibrium: Berthollet's Law of Mass Action." *Chymia*, 8: 105–45.
Holmes, Frederic L. 1964. "Introduction to Liebig's *Animal Chemistry*." New York: Johnson Reprint Corporation.
Holmes, Frederic L. 1971. "Analysis by Fire and Solvent Extractions: The Metamorphosis of a Tradition." *Isis*, 62: 128–48.
Holmes, Frederic L. 1989a. *Eighteenth Century Chemistry as an Investigative Enterprise*. Berkeley, CA: Office for History of Science and Technology.
Holmes, Frederic L. 1989b. "The Complementarity of Teaching and Research in Liebig's Laboratory." *Osiris*, 5: 121–64.
Holter, H., and Max K. Møller (eds). 1976. *The Carlsberg Laboratory 1876–1976*. Copenhagen: Rhodos Publishing House.
Homburg, Ernst. 1992. "The Emergence of Research Laboratories in the Dyestuffs Industry, 1870–1900." *The British Journal for the History of Science*, 25: 91–111.
Homburg, Ernst. 1998. "Two Factions, One Profession: The Chemical Profession in German Society 1780–1870." In David Knight and Helge Kragh (eds), *The Making of the Chemist: The Social History of Chemistry in Europe, 1789–1914*. Cambridge: Cambridge University Press.
Homburg, Ernst. 1999. "The Rise of Analytical Chemistry and Its Consequences for the Development of the German Chemical Profession (1780–1860)." *Ambix*, 46: 1–32.
Homburg, Ernst, Anthony S. Travis, and Harm G. Schröter (eds). 1998. *The Chemical Industry in Europe, 1850–1914: Industrial Growth, Pollution, and Professionalization*. Dordrecht: Springer.
Ihde, Aaron J. 1964. *The Development of Modern Chemistry*. New York: Harper & Row.
Jackson, Catherine M. 2006. "Re-Examining the Research School: August Wilhelm Hofmann and the Re-Creation of a Liebigian Research School in London." *History of Science*, 44: 281–319.
Jackson, Catherine M. 2008. "Visible Work: The Role of Students in the Creation of Liebig's Giessen Research School." *Notes and Records of the Royal Society of London*, 62: 31–49.
Jackson, Catherine M. 2011. "Chemistry as the Defining Science: Discipline and Training in Nineteenth-Century Chemical Laboratories." *Endeavour*, 35: 55–62.
Jackson, Catherine. 2014a. "The Curious Case of Coniine: Constructive Synthesis and Aromatic Structure Theory." In Carsten Reinhardt and Ursula Klein (eds), *Objects of Chemical Inquiry*. Sagamore Beach, MA: Science History Publications.
Jackson, Catherine M. 2014b. "Synthetical Experiments and Alkaloid Analogues: Liebig, Hofmann, and the Origins of Organic Synthesis." *Historical Studies in the Natural Sciences*, 44: 319–63.
Jackson, Catherine M. 2015a. "Chemical Identity Crisis: Glass and Glassblowing in the Identification of Organic Compounds." *Annals of Science*, 72: 187–205.
Jackson, Catherine M. 2015b. "The 'Wonderful Properties of Glass': Liebig's *Kaliapparat* and the Practice of Chemistry in Glass." *Isis*, 106: 43–69.

Jackson, Roland. 2018. *The Ascent of John Tyndall: Victorian Scientist, Mountaineer, and Public Intellectual*. New York: Oxford University Press.
Jago, William. 1886. *The Chemistry of Wheat, Flour, and Bread; and the Technology of Breadmaking*. London: Simpkin, Marshall, Hamilton, Kent, & Co.
James, Frank A.L. 2010. *Michael Faraday: A Very Short Introduction*. New York: Oxford University Press.
James, Frank A.L. (ed.). 2011. *Michael Faraday. The Chemical History of a Candle*. Oxford: Oxford University Press.
James, Frank A.L. 1995. "Science as a Cultural Ornament: Bunsen, Kirchhoff and Helmholtz in Mid-Nineteenth Century Baden." *Ambix*, 42: 1–9.
James, Frank A.L. 2000. *Guides to the Royal Institution of Great Britain: 1, History*. London: Royal Institution of Great Britain.
Jensen, William B. 1986. "The Development of Blowpipe Analysis." In John T. Stock and Mary Virginia Oma (eds), *The History and Preservation of Chemical Instrumentation*. Dordrecht: Springer.
Jensen, William B. 2005. "The Origin of the Bunsen Burner." *Journal of Chemical Education*, 82: 518.
Jensen, William B. 2006. "The Origin of the Term Allotrope." *Journal of Chemical Education*, 83: 838–9.
Jensen, William B. 2011. "Physical Chemistry before Ostwald: The Textbooks of Josiah Parsons Cooke." *Bulletin for the History of Chemistry*, 36: 10–21.
John G. Waite Associates, Architects PLLC. 2017. *The Rotunda Chemical Hearth: Historic Structure Report*. Charlottesville: The University of Virginia.
Johnson, Jeffrey A. 1985. "Academic Chemistry in Imperial Germany." *Isis*, 76: 500–24.
Johnson, Jeffrey A. 1989. "Hierarchy and Creativity in Chemistry, 1871–1914." *Osiris*, 2nd Series: 214–40.
Johnson, Jeffrey Allan. 2017. "Between Nationalism and Internationalism: The German Chemical Society in Comparative Perspective, 1867–1945." *Angewandte Chemie International Edition*, 56: 11044–58.
Jones, David. 1988. *The Origins of Civic Universities. Manchester, Leeds and Liverpool*. London: Routledge.
Kaji, Masanori. 2003. "Mendeleev's Discovery of the Periodic Law: The Origin and Reception." *Foundations of Chemistry*, 5: 189–214.
Kaji, Masanori, Helge Kragh, and Palló Gabor (eds). 2015. *Early Responses to the Periodic System*. Oxford: Oxford University Press.
Kargon, Robert. 1977. *Science in Victorian Manchester: Enterprise and Expertise*. Baltimore, MD: Johns Hopkins University Press.
Kauffman, George B. 1966. *Alfred Werner: Founder of Coordination Chemistry*. Berlin: Springer.
Kauffman, George B. 1968. *Classics in Coordination Chemistry. Part 1: The Selected Papers of Alfred Werner*. New York: Dover.
Kay-Williams, Susan. 2013. *The Story of Colour in Textiles; Imperial Purple to Denim Blue*. London: Bloomsbury.
Kedrov, B.M. 1981. "Mendeleev, Dimitry Ivanovich." In Charles Gillespie (ed.), *Dictionary of Scientific Biography*. New York: Scribner's.
Keene, Melanie. 2013. "From Candles to Cabinets: 'Familiar Chemistry' in Early Victorian Britain." *Ambix*, 60: 54–77.
Kekulé, August. 1861. *Lehrbuch der organischen Chemie*. Erlangen: Enke.

Kekulé, August. 1867. *Chemie der Benzolderivate oder der aromatische Substanzen*. Erlangen: Enke.

Kelly, Matt. 2015. "Jeffersonian-Era Chemistry Hearth Preserved in Rotunda Wall." *UVA Today*.

Kikuchi, Yoshiyuki. 2013. *Anglo-American Connections in Japanese Chemistry: The Lab as Contact Zone*. New York: Palgrave-Macmillan.

Kilburn, Matthew. 2009. "Gibbs, William." Oxford: Oxford Dictionary of National Biography.

Kim, Mi Gyung. 2006. "Wilhelm Ostwald (1853–1932)." *HYLE: International Journal for the Philosophy of Chemistry*, 12: 141–8.

King, M. Christine. 1981. "Experiments With Time: Progress and Problems in the Development of Chemical Kinetics." *Ambix*, 28: 70–82.

King, M. Christine. 1984. "The Course of Chemical Change: The Life and Times of Augustus G. Vernon Harcourt (1834–1919)." *Ambix*, 31: 16–31.

Kirchhoff, Gustav, and Robert Bunsen. 1860. "Chemical Analysis by Spectrum-Observations." *London, Edinburgh, and Dublin Philosophical Magazine and Journal of Science*, 22: 889–109.

Klein, Ursula. 2001. "The Creative Power of Paper Tools in Early Nineteenth Century Chemistry." In Ursula Klein (ed.), *Tools and Modes of Representation in the Laboratory Sciences*. Dordrecht: Kluwer.

Klein, Ursula. 2003. *Experiments, Models, Paper Tools: Cultures of Organic Chemistry in the Nineteenth Century*. Stanford, CA: Stanford University Press.

Klein, Ursula. 2008. "The Laboratory Challenge: Some Revisions of the Standard View of Early Modern Experimentation." *Isis*, 99: 769–82.

Klein, Ursula, and Wolfgang Lefèvre. 2007. *Materials in Eighteenth-Century Science: A Historical Ontology*. Cambridge, MA: MIT Press.

Knight, David M. 1970. *Classical Scientific Papers: Chemistry, Second Series. Papers on the Nature and Arrangement of the Chemical Elements*. London/New York: Mills & Boon American Elsevier.

Knight, David M. 1986. *The Age of Science: The Scientific World-View in the Nineteenth Century*. Oxford: Blackwell.

Kohler, Robert. 1971. "The Background to Eduard Buchner's Discovery of Cell-Free Fermentation." *Journal of the History of Biology*, 4: 35–61.

Kohler, Robert E. 1973. "The Enzyme Theory and the Origin of Biochemistry." *Isis*, 64: 181–96.

Kohn, M. 1950. "Remarks on the History of Laboratory Burners." *Journal of Chemical Education*, 27: 514–16.

Kragh, Helge. 1984. "Julius Thomsen and Classical Thermochemistry." *British Journal for the History of Science*, 17: 255–72.

Kragh, Helge. 2016. *Julius Thomsen: A Life in Chemistry and Beyond*. Copenhagen: Royal Danish Academy of Sciences and Letters.

Krätz, Otto. 1990. "Chemie im Spiegel der Schöngeistigen Literatur zur Zeit Leopold Gmelins." In W. Lippert (ed.), *Der 200. Geburtstag von Leopold Gmelin*. Frankfurt/M.: Gmelin Institut.

Kremer, Richard L. 2009. "Physiology." In Peter J. Bowler and J.V. Pickstone (eds), *The Modern Biological and Earth Sciences*, in Cambridge History of Science, vol 6, David C. Lindberg and Ronald L. Numbers (eds), Cambridge: Cambridge University Press.

Kumar, Prakash. 2012. *Indigo Plantations and Science in Colonial India*. Cambridge: Cambridge University Press.

Kutney, Gerald. 2013. *Sulfur: History, Technology, Applications and Industry*. Toronto: ChemTech Publishing.

Labinger, Jay. 2011. "Chemistry." In Bruce Clarke and Manuela Rossini (eds), *The Routledge Companion to Literature and Science*. New York: Routledge.

Laidler, Keith J. 1985. "Chemical Kinetics and the Origins of Physical Chemistry." *Archive for History of Exact Sciences*, 32: 43–75.

Laidler, Keith J. 1993. *The World of Physical Chemistry*. Oxford: Oxford University Press.

Landry, Jennifer. 2014. "Cabinets for the Curious." *Distillations*. Available online: https://www.sciencehistory.org/distillations/magazine/cabinets-for-the-curious (accessed May 17, 2021).

Lane, Joan. 2001. *A Social History of Medicine. Health, Healing and Disease in England, 1750–1950*. London: Routledge.

Lankford, John. 1984. "The Impact of Photography on Astronomy." In Owen Gingerich (ed.), *Astrophysics and Twentieth Century Astronomy to 1950*. Cambridge: Cambridge University Press.

Latour, Bruno. 1993. *The Pasteurization of France*. Cambridge, MA: Harvard University Press.

Lavine, Matthew. 2013. *The First Atomic Age: Scientists, Radiations, and the American Public, 1895–1945*. New York: Palgrave MacMillan.

Lavoisier, Antoine. 1789. *Elements of Chemistry*. New York: Dover [1965].

Lavoisier, Antoine, and Pierre-Simon La Place. 1783. *Memoir on Heat*. New York: Watson Academic Publications [1982].

Le Roux, Muriel. 2015. "From Science to Industry: The Sites of Aluminium in France from the Nineteenth to the Twentieth Century." *Ambix*, 62: 114–37.

Le Roux, Thomas, and Jean-Baptiste Fressoz. 2011. "Protecting Industry and Commodifying the Environment: The Great Transformation of French Pollution Regulation, 1700–1840." In Genevieve Guilbaud and Stephen Mosley (eds), *Common Ground. Integrating the Social and Environmental in History*. Newcastle (UK): Cambridge Scholars Publishing.

Leicester, Henry M. 1951. "Germain Henri Hess and the Foundations of Thermochemistry." *Journal of Chemical Education*, 28: 581.

Leigh, G. Jeffery, and Alan J. Rocke. 2016. "Women and Chemistry in Regency England: New Light on the Marcet Circle." *Ambix*, 63: 28–45.

Leigh, G.J. 2004. *The World's Greatest Fix: A History of Nitrogen and Agriculture*. Oxford: Oxford University Press.

Lesch, John E. 1981. "Conceptual Change in an Empirical Science: The Discovery of the First Alkaloids." *Historical Studies in the Physical Sciences*, 11: 305–28.

Lesch, John E. 1984. *Science and Medicine in France: The Emergence of Experimental Physiology, 1790–1855*. Cambridge, MA: Harvard University Press.

Levere, Trevor H. 1971. *Affinity and Matter: Elements of Chemical Philosophy, 1800–1865*. Oxford: Clarendon.

Levere, Trevor H. 1975. "Arrangement and Structure: A Distinction and a Difference." In O.B. Ramsay (ed.), *Van't Hoff–Le Bel Centennial*. Washington, DC: American Chemical Society.

Levere, Trevor H. 2000. "Measuring Gases and Measuring Goodness." In Frederic L. Holmes and Trevor H. Levere (eds), *Instruments and Experimentation in the History of Chemistry*. Cambridge, MA: MIT Press.

Levere, Trevor H. 2001. *Transforming Matter: A History of Chemistry From Alchemy to the Buckyball*. Baltimore, MD: Johns Hopkins University Press.
Levere, Trevor H. 2010. "Sons of Genius: Chemical Manipulation and its Shifting Norms From Joseph Black to Michael Faraday." *Bulletin for the History of Chemistry*, 35: 1–6.
Lewis, David E. 1994a. "The University of Kazan—Provincial Cradle of Russian Organic Chemistry: Part I. Nikolai Zinin and the Butlerov School." *Journal of Chemical Education*, 71: 39–42.
Lewis, David E. 1994b. "The University of Kazan—Provincial Cradle of Russian Organic Chemistry: Part II: Aleksandr Zaitzev and His Students." *Journal of Chemical Education*, 71: 93–7.
Lewis, David E. 2012. *Early Russian Chemists and their Legacy*. New York: Springer.
Lewis, Gilbert N. 1923. *Valence and the Structure of Atoms and Molecules*. New York: Chemical Catalog Company.
Liebig, Justus. 1847. *Chemistry, and its Application to Physiology, Agriculture, and Commerce*. New York: Fowlers and Wells.
Lindee, M. Susan. 1991. "The American Career of Jane Marcet's *Conversations on Chemistry*, 1806–1853." *Isis*, 82: 8–23.
Linden, Stanton J. 1996. *Darke Hieroglyphicks. Alchemy in English Literature From Chaucer to the Restoration*. Lexington, KY: University Press of Kentucky.
Lodge, F.S. 1938. "Potash in the Fertilizer Industry." *Industrial and Engineering Chemistry*, 30: 878–82.
Longair, Malcolm S. 2006. "*The Cosmic Century: A History of Astrophysics and Cosmology*." Cambridge: Cambridge University Press.
Lucier, Paul. 2008. *Scientists and Swindlers: Consulting on Coal and Oil in America, 1820–1890*. Baltimore, MD: Johns Hopkins University Press.
Lunge, Georg. 1886. *Sulphuric Acid and Alkali,* Vol. 2. London: John Van Voort.
MacLeod, Roy M. 1965. "The Alkali Acts Administration, 1863–1884: The Emergence of the Civil Scientist." *Victorian Studies*, 9: 85–112.
MacLeod, Roy M., Russell G. Egdell, and Elizabeth Bruton (eds). 2018. *For Science, King and Country: The Life and Legacy of Henry Moseley*. London: Uniform Publishing Group.
Malley, Marjorie. 1979. "The Discovery of Atomic Transmutation: Scientific Styles and Philosophies in France and Britain." *Isis*, 70: 213–23.
Malley, Marjorie C. 2011. *Radioactivity: A History of a Mysterious Science*. New York: Oxford University Press.
Marcet, Jane. 1817. *Conversations on Chemistry*. London: Longman, Hurst, Rees, Orme and Brown.
Marsh, J. 1836. "Account of a Method of Separating Small Quantities of Arsenic from Substances with which it may be Mixed." *Edinburgh New Philosophical Journal*, 21: 229–36.
Mauskopf, Seymour H. 1976. *Crystals and Compounds: Molecular Structure and Composition in Nineteenth Century French Science*. Philadelphia, PA: Transactions of the American Philosophical Society.
Mauskopf, Seymour. 1999. "'From an Instrument of War to an Instrument of the Laboratory: The Affinities Certainly Do Not Change' Chemists and the Development of Munitions, 1785–1885." *Bulletin for the History of Chemistry*, 24: 1–15.
Mawe, John. 1825. *Instructions for the Use of the Blow-Pipe, and Chemical Tests, with Additions and Observations Derived from the Recent Publication of Professor Berzelius*. London: Longman.

McEvoy, John G. 1988. "Continuity and Discontinuity in the Chemical Revolution." *Osiris*, 4: 195–213.

McGucken, William. 1969. *Nineteenth Century Spectroscopy: Development of the Understanding of Spectra, 1802–1897*. Baltimore, MD: Johns Hopkins University Press.

Meadows, Jack. 2004. *The Victorian Scientist. The Growth of a Profession*. London: The British Library.

Meakin, David. 1995. *Hermetic Fictions: Alchemy and Irony in the Novel*. Keele: Keele University Press.

Meinel, Christoph. 2004. "Molecules and Croquet Balls." In Soraya De Chadaverian and Nick Hopwood (eds), *Models: The Third Dimension of Science*. Stanford, CA: Stanford University Press.

Melhado, Evan M. 1980. "Mitscherlich's Discovery of Isomorphism." *Historical Studies in the Physical Sciences*, 11: 87–123.

Melillo, Edward D. 2012. "The First Green Revolution: Debt Peonage and the Making of the Nitrogen Fertilizer Trade, 1840–1930." *American Historical Review*, 117: 1028–60.

Mertens, Joost. 2019. "Schweinfurt Green and the Sanitary Police: The Fight against Copper Arsenite Pigments." In Elisabeth Vaupel and Ernst Homburg (eds), *Hazardous Chemicals: Agents of Risk and Change (1800–2000)*. New York: Berghahn.

Meyer-Thurow, Georg. 1982. "The Industrialization of Invention: A Case Study from the German Chemical Industry." *Isis*, 73: 363–81.

Miles, William D. 1970. "William James MacNeven and Early Laboratory Instruction in the United States." *Ambix*, 17: 143–52.

Miller, James. 2006. *Fertile Fortune: The Story of Tyntesfield*. London: National Trust.

Miller, William A. 1856. *Elements of Chemistry: Theoretical and Practical*. London: J. W. Parker and Son.

Molony, Barbara. 1990. *Technology and Investment: The Prewar Japanese Chemical Industry*. Cambridge, MA: Harvard University Press.

Morrell, J.B. 1972. "The Chemist Breeders: The Research Research Schools of Liebig and Thomson." *Ambix*, 19: 1–46.

Morrell, Jack. 1993. "W. H. Perkin Jr., at Manchester and Oxford: From Irwell to Isis." *Osiris*, 8: 104–26.

Morrell, Jack, and Arnold Thackray. 1984. *Gentlemen of Science: Early Correspondence of the British Association for the Advancement of Science*. London: Royal Historical Society.

Morris, Peter J.T. 1989. "The Legacy of Ludwig Mond." *Endeavour*, 13: 34–40.

Morris, Peter J.T. (ed.). 2002. *From Classical to Modern Chemistry: The Instrumental Revolution*. Philadelphia, PA: Chemical Heritage Foundation.

Morris, Peter J.T. 2003. *Chemical Industry Before 1850*. Oxford: Oxford University Press.

Morris, Peter J.T. 2015. *The Matter Factory: A History of the Chemistry Laboratory*. London: Reaktion Books.

Morris, Peter J.T., W.A. Campbell, and H. L. Roberts (eds). 1991. *Milestones in 150 Years of the Chemical Industry*. Cambridge: Royal Society of Chemistry.

Morris, Peter J.T., Colin A. Russell, and John Graham Smith (eds). 1988. *Archives of the British Chemical Industry 1750-1914. A Handlist*. Oxfordshire: British Society for the History of Science.

Morrisson, Mark S. 2007. *Modern Alchemy: Occultism and the Emergence of Atomic Theory*. New York: Oxford University Press.
Munday, Pat. 2000. "Eben N. Horsford." *American National Biography Online*.
Muspratt, James Sheridan. 1860. *Chemistry, Theoretical, Practical and Analytical as Applied and Related to Arts and Manufactures*. 2 vols. Glasgow: William MacKenzie.
Nawa, Christine. 2014. "A Refuge for Inorganic Chemistry: Bunsen's Heidelberg Laboratory." *Ambix*, 61: 115–40.
Newell, Edmund. 1990. "'Copperopolis': The Rise and Fall of the Copper Industry in in the Swansea District, 1826–1821." *Business History*, 32: 75–97.
Newell, Edmund. 1997. "Atmospheric Pollution and the British Copper Industry, 1690–1920." *Technology and Culture*, 38: 655–89.
Nicholson, Peter. 1845. *The Builder and Workman's New Director*. London: A. Fullerton and Co.
Nielsen, Anita Kildebæk, and Sona Strbánová. 2008. *Creating Networks in Chemistry: The Founding and Early History of Chemical Societies in Europe*. Cambridge: Royal Society of Chemistry.
Nieto-Galan, Agustí. 1997. "Calico Printing and Chemical Knowledge in Lancashire in the Early Nineteenth Century: The Life and 'Colours' of John Mercer." *Annals of Science*, 54: 1–28.
Nieto-Galan, Agustí. 2001. *Colouring Textiles: A History of Natural Dyestuffs in Industrial Europe*. Dordrecht: Kluwer.
Nieto-Galan, Agustí. 2016. *Science in the Public Sphere: A History of Lay Knowledge and Expertise*. London: Routledge.
Nye, Mary Jo. 1984. *The Question of the Atom: From the Karlsruhe Congress to the First Solvay Conference, 1860–1911*. Los Angeles, CA: Tomash Publishers.
Nye, Mary Jo. 1986. *Science in the Provinces: Scientific Communities and Provincial Leadership in France, 1860–1930*. Berkeley: University of California Press.
Nye, Mary Jo. 1993. *From Chemical Philosophy to Theoretical Chemistry: Dynamics of Matter and Dynamics of Disciplines, 1800–1950*. Berkeley: University of California Press.
Nye, Mary Jo. 1996. *Before Big Science: The Pursuit of Modern Chemistry and Physics*. New York: Twayne Publishers.
O'Dea, W.T. 1964. *Making Fire*. London: HMSO.
Obrist, Barbara. 1996. "Art et nature dans l'alchimie médiévale." *Revue d'Histoire des Sciences*, 49: 215–86.
Ogrinc, W. H. L. 1980. "Western Society and Alchemy From 1200 to 1500." *Journal of Medieval History*, 6: 103–33.
Ostwald, Wilhelm. 1905. *Conversations on Chemistry, First Steps in Chemistry, Part I: General Chemistry*. New York: John Wiley & Sons.
Ostwald, Wilhelm. 1912. *Outlines of General Chemistry*. London: MacMillan and Co.
Palladino, Paolo. 1990. "Stereochemistry and the Nature of Life: Mechanist, Vitalist, and Evolutionary Perspectives." *Isis*, 81: 44–67.
Pancaldi, Guiliano. 2003. *Volta: Science and Culture in the Age of Enlightenment*. Princeton, NJ: Princeton University Press.
Partington, J.R. 1961–1970. *A History of Chemistry*. 4 vols. London: MacMillan.
Picard, Liza. 2005. *The Commodity Culture of Victorian England*. Stanford, CA: Stanford University Press.
Principe, Lawrence. 2012. *The Secrets of Alchemy*. Chicago, IL: University of Chicago Press.

Raitt, Gordon J. 1966. *Modern Chemistry. Applied and Social Aspects.* London: Edward Arnold.
Ramberg, Peter J. 2000. "The Death of Vitalism and the Birth of Organic Chemistry: Wöhler's Urea Synthesis in Textbooks of Organic Chemistry." *Ambix*, 47: 170–95.
Ramberg, Peter J. 2003. *Chemical Structure, Spatial Arrangement: The Early History of Stereochemistry, 1874–1914.* Aldershot: Ashgate.
Ramberg, Peter J. 2014. "Partial Valence, Residual Affinity, and Early Stereochemistry." In Carsten Reinhardt and Ursula Klein (eds), *Objects of Chemical Inquiry*. Sagamore Beach, MA: Science History Publications.
Ramberg, Peter J. 2015. "Chemical Research and Instruction in Zürich, 1833–1872." *Annals of Science*, 72: 170–86.
Ramberg, Peter J., and Geert J. Somsen. 2001. "The Young J. H. Van 't Hoff: The Background to the Publication of his 1874 Pamphlet on the Tetrahedral Carbon Atom, Together with a New English Translation." *Annals of Science*, 58: 51–74.
Rayner-Canham, Marelene F., and Geoffrey Rayner-Canham. 1998. *Women in Chemistry: Their Changing Role From Alchemical Times to the Mid-Twentieth Century.* Philadelphia, PA: American Chemical Society.
Read, John. 1947. *The Alchemist in Life, Literature and Art.* London: Nelson.
Reader, William J. 1970. *Imperial Chemical Industries: A History, Vol I: The Forerunners, 1870–1926.* Oxford: Oxford University Press.
Reed, Peter. 2008. "Acid Towers and the Control of Chemical Pollution 1823–1876." *Transactions of the Newcomen Society*, 78: 99–126.
Reed, Peter. 2012. "The Alkali Inspectorate 1874–1906: Pressure for Wider and Tighter Pollution Regulation." *Ambix*, 59: 131–51.
Reed, Peter. 2014. *Acid Rain and the Rise of the Environmental Chemist in Nineteenth-Century Britain.* Burlington, VT: Ashgate.
Reed, Peter. 2015. *Entrepreneurial Ventures in Chemistry: The Muspratts of Liverpool, 1793–1934.* Burlington, VT: Ashgate.
Reed, Peter. 2017. "John Fletcher Moulton and the Transforming Aftermath of the Chemists' War." *The International Journal for the History of Engineering and Technology*, 87: 1–19.
Reinhardt, Carsten, and Anthony S. Travis. 2000. *Heinrich Caro and the Creation of Modern Chemical Industry.* Dordrecht/Boston: Kluwer.
"Report of the Select Committee on the Adulteration of Food, Drink and Drugs." *The Times*, 20 August, 1856.
Rezneck, Samuel. 1970. "The European Education of an American Chemist and Its Influence on 19th-Century America: Eben Norton Horsford." *Technology and Culture*, 11: 366–88.
Richards, Edgar. 1890. "Legislation on Food Adulteration." *Science*, 16: 101–4.
Richards, Thomas. 1990. *The Commodity Culture of Victorian England. Advertising and Spectacle, 1851–1914.* Stanford, CA: Stanford University Press.
Roberts, Gerrylynn K. 1976a. *The Royal College of Chemistry (1845–1853): A Social History of Chemistry in Early-Victorian England.* Ph.D. thesis, The Johns Hopkins University.
Roberts, Gerrylynn K. 1976b. "The Establishment of the Royal College of Chemistry: An Investigation of the Social Context of Early-Victorian Chemistry." *Historical Studies in the Physical Sciences*, 7: 437–85.
Rocke, Alan J. 1978. "Atoms and Equivalents: The Early Development of the Chemical Atomic Theory." *Historical Studies in the Physical Sciences*, 9: 225–63.

Rocke, Alan J. 1981. "Kekulé, Butlerov, and the Historiography of the Theory of Chemical Structure." *British Journal for the History of Science*, 14: 27–57.

Rocke, Alan J. 1984. *Chemical Atomism in the Nineteenth Century: From Dalton to Canizarro*. Columbus: Ohio State University Press.

Rocke, Alan J. 1985. "Hypothesis and Experiment in the Early Development of Kekulé's Benzene Theory." *Annals of Science*, 42: 355–81.

Rocke, Alan J. 1990. "Methodology and its Rhetoric in Nineteenth Century Chemistry: Induction Versus Hypothesis." In Elizabeth Garber (ed.), *Beyond History of Science: Essays in Honor of Robert E. Schofield*. Bethelem, PA: Lehigh University Press.

Rocke, Alan J. 1993. *The Quiet Revolution: Hermann Kolbe and the Science of Organic Chemistry*. Berkeley: University of California Press.

Rocke, Alan J. 1994. "History and Science, History of Science: Adolphe Wurtz and the Renovation of the Academic Professions in France." *Ambix*, 41: 20–32.

Rocke, Alan J. 2000a. *Nationalizing Science: Adolphe Wurtz and the Battle for French Chemistry*. Cambridge, MA: MIT Press.

Rocke, Alan J. 2000b. "Organic Analysis in Comparative Perspective." In Frederic L. Holmes and Trevor H. Levere (eds), *Instruments and Experimentation in the History of Chemistry*. Cambridge, MA: MIT Press.

Rocke, Alan J. 2003. "Origins and Spread of the 'Giessen Model' in University Science." *Ambix*, 50: 90–115.

Rocke, Alan J. 2010. *Image and Reality: Kekulé, Kopp, and the Scientific Imagination*. Chicago, IL: The University of Chicago Press.

Rocke, Alan J. 2018a. "Theory versus Practice in German Chemistry: Erlenmeyer Beyond the Flask." *Isis*, 109: 254–75.

Rocke, Alan J. 2018b. "Ideas in Chemistry: The Pure and the Impure." *Isis*, 109: 577–86.

Rocke, Alan J., and Hermann Kopp. 2012. *From the Molecular World: A Nineteenth-Century Science Fantasy*. New York: Springer.

Roderick, Gordon W., and Michael D. Stephens. 1972. *Scientific and Technical Education in Nineteenth-Century England*. Newton Abbott: David and Charles.

Romer, Alfred (ed.). 1964. *The Discovery of Radioactivity and Transmutation*. New York: Dover.

Roscoe, Henry. 1906. *The Life & Experiences of Sir Henry Enfield Roscoe*. London: Macmillan.

Rossiter, Margaret W. 1975. *The Emergence of Agricultural Science: Justus Liebig and the Americans, 1840–1880*. New Haven, CT: Yale University Press.

Rossotti, Hazel. 2006. *Chemistry in the Schoolroom: 1806: Selections From Mrs. Marcet's* Conversations on Chemistry. Bloomington, IN: AuthorHouse.

Rowlinson, John S. 2009. "Chemistry Comes of Age: The 19th Century." In R.J.P. Williams, A. Chapman, and J.S. Rowlinson (eds), *Chemistry at Oxford: A History from 1600 to 2005*. Cambridge: Royal Society of Chemistry.

Rowlinson, P.J. 1982. "Food Adulterations: Its Control in 19th Century Britain." *Interdisciplinary Science Review*, 7: 63–72.

Russell, Colin A. 1971. *The History of Valency*. New York: Humanities Press.

Russell, Colin A. 1987. "The Changing Role of Synthesis in Organic Chemistry." *Ambix*, 34: 168–80.

Russell, Colin A. 1996. *Edward Frankland: Chemistry, Controversy and Conspiracy in Victorian England*. Cambridge: Cambridge University Press.

Russell, Colin A. 2000a. "Chemical Techniques in a Pre-Electronic Age: The Remarkable Apparatus of Edward Frankland." In Frederic L. Holmes and Trevor Levere (eds), *Instruments and Experimentation in the History of Chemistry*. Cambridge, MA: MIT Press.

Russell, Colin A. (ed.). 2000b. *Chemistry, Society and Environment: A New History of the British Chemical Industry*. Cambridge: Royal Society of Chemistry.

Russell, Colin A., Noel Coley, and Gerrylynn K. Roberts. 1977. *Chemists by Profession. The Origins and Rise of the Royal Institute of Chemistry*. Milton Keynes: The Open University Press.

Russell, Colin A., and John A. Hudson. 2012. *Early Railway Chemistry and Its Legacy*. Cambridge: Royal Society of Chemistry.

Rutherford, Ernest. 1906. *Radioactive Transformations*. London: Constable.

Sagatos, O. J. 2005. *Progress of a Different Nature: Hydro 1905–2005*. Oslo: Pax Forlag.

Sanderson, Michael. 1975. *The Universities in the Nineteenth Century*. London: Routledge.

Scerri, Eric R. 2007. *The Periodic Table: Its Story and its Significance*. New York: Oxford University Press.

Schelar, Virginia M. 1966. "Thermochemistry and the Third Law of Thermodynamics." *Chymia*, 11: 99–124.

Schiff, Eric. 1971. *Industrialization Without National Patents: The Netherlands, 1862–1912; Switzerland, 1887–1907*. Princeton, NJ: Princeton University Press.

Schummer, Joachim. 1997. "Scientometric Studies on Chemistry I: The Exponential Growth of Chemical Substances, 1800–1995." *Scientometrics*, 39: 107–23.

Schummer, Joachim. 2006. "Historical Roots of the 'Mad Scientist': Chemists in Nineteenth-Century Literature." *Ambix*, 53: 99–127.

Schummer, Joachim. 2008. "Frankenstein und die literarische Figur des verrückten Wissenschaftlers." In Betsy van Schlun and Michael Neumann (eds), *Mythen Europas: Schlüsselfiguren der Imagination*. Regensburg: Pustet.

Schummer, Joachim. 2022. "Art and Representation." In Peter J.T. Morris (ed.), *The Bloomsbury Cultural History of Chemistry in the Modern Age*. London: Bloomsbury.

Schummer, Joachim, Bernadette Bensaude-Vincent, and Brigitte van Tiggelen (eds). 2007. *Public Image of Chemistry*. Singapore: World Scientific Publishing.

Schummer, Joachim, and Tami I. Spector. 2007. "The Visual Image of Chemistry: Perspectives from the History of Art and Science." *HYLE: International Journal for Philosophy of Chemistry*, 13: 3–41.

Schurer, H. 1951. "The Macintosh: The Paternity of an Invention." *Transactions of the Newcomen Society*, 28: 77–87.

Sepper, Dennis L. 1990. "Goethe, Colour and the Science of Seeing." In Andrew Cunningham and Nicholas Jardine (eds), *Romanticism in the Sciences*. Cambridge: Cambridge University Press.

Servos, John W. 1990. *Physical Chemistry From Ostwald to Pauling: The Making of a Science in America*. Princeton, NJ: Princeton University Press.

Shelley, Mary. 1831. *Frankenstein, Or, the Modern Prometheus*. London: Henry Colburn and Richard Bentley.

Sherard, Robert H. 1896. "The White Slaves of England." *Pearson's Magazine II*, 48–55.

Sherman, Irwin W. 2017. *Drugs That Changed the World: How Therapeutic Agents Shaped Our Lives*. Boca Raton, FL: CRC Press.

Silliman Jr., Benjamin. 1856. *First Principles of Chemistry, for the Use of Colleges and Schools*. Philadelphia, PA: H.C. Peck & Theo. Bliss.

Smeaton, W.A. 1954. "The Early History of Laboratory Instruction in Chemistry at the École Polytechnique, Paris and Elsewhere." *Annals of Science*, 10: 224–33.

Smil, Vaclav. 2001. *Enriching the Earth: Fritz Haber, Carl Bosch, and the Transformation of World Food Production*. Cambridge, MA: MIT Press.

Smith, John G. 1979. *The Origins and Early Development of the Heavy Chemical Industry in France*. Oxford: Clarendon Press.

Sneader, Walter. 2005. *A History of Drug Discovery*. New York: Wiley.

Sogner, Knut. 2014. *Creative Power. Elkem 110 Years, 1904–2014*. Oslo: Elkem.

Sonntag, Otto. 1977. "Religion and Science in the Thought of Liebig." *Ambix*, 24: 159–69.

Spronsen, J.W. van. 1969. *The Periodic System of the Chemical Elements: A History of the First Hundred Years*. Amsterdam: Elsevier.

Stolzenberg, Dietrich. 2004. *Fritz Haber: Chemist, Nobel Laureate, German, Jew*. Philadelphia, PA: Chemical Heritage Foundation.

Stookey, Byron. 1965. "William James MacNeven (1763–1841): Versatile Professor in New York's College of Physicians and Surgeons." *Bulletin of the New York Academy of Medicine*, 41: 1037–51.

Stranges, Anthony. 1982. *Electrons and Valence*. College Station, TX: Texas A&M University Press.

Szabadváry, Ferenc. 1966. *History of Analytical Chemistry*. London: Pergamon.

Tarr, Joel A. 1996. *The Search for the Ultimate Sink: Urban Pollution in Historical Perspective*. Akron, OH: University of Akron Press.

Thackray, Arnold et al. 1985. *Chemistry in America, 1876–1976: Historical Indicators*. Dordrecht: Reidel.

"The Society of Chemical Industry." July 1931. *Journal of the Society of Chemical Industry (Jubilee Edition)*, 9–17.

Tower, Donald B. 1994. *Brain Chemistry and the French Connection, 1791–1841: An Account of the Chemical Analyses of the Human Brain by Thouret (1791), Fourcroy (1793), Vauquelin (1811), Couerbe (1834), and Fremy (1841)*. New York: Raven Press.

Travis, Anthony S. 1989. "Science as a Receptor of Technology: Paul Ehrlich and the Synthetic Dyestuffs Industry." *Science in Context*, 3: 383–408.

Travis, Anthony S. 1993. *The Rainbow Makers: The Origins of the Synthetic Dyestuffs Industry in Western Europe*. Bethlehem, PA: Lehigh University Press.

Travis, Anthony S. 2008. "Models for Biomedical Research: The Theory and Practice of Paul Ehrlich." *History and Philosophy of the Life Sciences*, 30: 79–98.

Travis, Anthony S. 2018. *Nitrogen Capture: The Growth of an International Industry (1900–1940)*. Cham: Springer.

Trescott, Martha M. 1989. *The Rise of the American Electrochemicals Industry, 1880–1910. Contributions in Economics and Economic History*. Westport, CT: Greenwood Press.

Tropp, Martin. 1976. *Mary Shelley's Monster*. Boston, MA: Houghton Mifflin.

Turner, Gerard L'Estrange. 1983. *Nineteenth-Century Scientific Instruments*. Berkeley: University of California Press.

Uekoetter, Frank. 2009. *The Smoke Age: Environmental Policy in Germany and the United States, 1880–1970*. Pittsburgh, PA: University of Pittsburgh Press.

Usselman, Melvyn C. et al. 2005. "Restaging Liebig: A Study in the Replication of Experiments." *Annals of Science*, 62: 1–55.
Usselman, Melvyn C. 2015. *Pure Intelligence: The Life of William Hyde Wallaston*. Chicago, IL: University of Chicago Press.
Warner, Deborah J. 2011. *Sweet Stuff: An American History of Sweeteners from Sugar to Sucralose*. Lanham, MD: Rowman & Littlefield.
Warren, Kenneth. 1988. *Chemical Foundations: The Alkali Industry in Britain until 1926*. Oxford: Clarendon Press.
Watson, Katherine D. 1995. "The Chemist as Expert: The Consulting Career of Sir William Ramsay." *Ambix*, 42: 143–59.
Watson, Katherine D. 2006. "Medical and Chemical Expertise in English Trials for Criminal Poisoning, 1750–1914." *Medical History*, 50: 373–90.
Watson, Katherine D. 2010. *Forensic Medicine in Western Society: A History*. New York: Routledge.
Weeks, Mary Elvira. 1968. *The Discovery of the Elements*. Eadston, PA: Journal of Chemical Education.
Werrett, Simon. 2014. "The Science of Destruction: Terrorism and Technology in the Nineteenth Century." In Carola Dietze and Claudia Verhoeven (eds), *The Oxford Handbook of the History of Terrorism*. Oxford: Oxford University Press.
Whitehead, Don. 1968. *The Dow Story: The History of the Dow Chemical Company*. New York: McGraw-Hill.
Whorton, James C. 2010. *The Arsenic Century: How Victorian Britain Was Poisoned at Home, Work and Play*. Oxford: Oxford University Press.
Williams, William D. 1992. "Some Early Chemical Slide Rules." *Bulletin for the History of Chemistry*, 12: 24–49.
Wilmot, Sarah. 1998. "Pollution and Public Concern: The Response of the Chemical Industry in Britain to Emerging Environmental Issues, 1860–1901." In Ernst Homburg, Anthony S. Travis, and Harm G. Schröter (eds), *The Chemical Industry in Europe, 1850–1914: Industrial Growth, Pollution, and Professionalization*. Dordrecht: Springer.
Wilson, Charles. 1970. *The History of Unilever*. London: Cessell & Co.
Wollaston, W.H. 1814. *A Synoptic Scale of Chemical Equivalents*. London: Bulmer.
Young, James Harvey. 1989. *Pure Food: Securing the Federal Food and Drugs Act of 1906*. Princeton, NJ: Princeton University Press.
Zapffe, Carl A. 1969. "Gustavus Hinrichs, Precursor of Mendeleev." *Isis*, 60: 461–76.
Ziolkowski, Theodore. 2015. *The Alchemist in Literature: From Dante to the Present*. New York: Oxford University Press.

LIST OF CONTRIBUTORS

Amy A. Fisher is Associate Professor of Science, Technology and Society at the University of Puget Sound, Tacoma, Washington, USA.

Yoshiyuki Kikuchi is Associate Professor at the Department of British and American Studies, Aichi Prefectural University, Nagakute, Japan.

Trevor Levere is University Professor of History of Science Emeritus at the University of Toronto, Canada.

Agustí Nieto-Galan is Professor of History of Science at the Institut d'Història de la Ciència (IHC), Universitat Autònoma de Barcelona, Spain.

Peter J. Ramberg is Professor of History of Science at Truman State University, Kirksville, Missouri, USA.

Peter Reed is an Independent Scholar, in Carmichael, California, USA.

Joachim Schummer is an Independent Scholar in Berlin, Germany.

Anthony S. Travis is Former Deputy Director, Sidney M. Edelstein Center, Hebrew University of Jerusalem, Israel.

INDEX

Abegg, Richard 34, 63–4
Abel, Frederick 142, 186
Abitur qualification 144, 195, 201
"absent-minded scientist" image 229–30
academic culture 22, 72–3
Accum, Friedrich 150
acetylene 147, 184
Acheson, Edward Goodrich 184
acid radicals 46, 50
acid rain 157, 160
"acid tower" 158–60
Acland, Henry 99
adulteration of foodstuffs 150–3
AGFA (company) 175, 176, 188
Agrippa of Nettlesheim 223
affinity, chemical 15–16, 42, 45, 60–3, 65, 103
air pollution 161
alchemical revival 36–7
Albert the Great 223
alchemy 36–7, 132, 218–22, 236
alcohol 121, 124
alizarin 27, 83, 175, 176
alkali industry 158–60, 165, 169–72, 177
alkaloids 14
Allhusen, Christian 170
allotropy 47, 48
aluminum 30
Aluminum Company of America (ALCOA) 30
Amagat, Emile 114

American Association for the Advancement of Science (AAAS) 212
American Chemical Society (ACS) 213
American Cyanamid Company 185
analysis 49, 67–76
 of organic compounds 49, 73, 110
analytical chemistry 15, 18, 211
Andersen, Hans Christian 223
anesthesia 14, 154–5, 178
aniline 174
Animal Chemistry 120, 125
apparatus and instrumentation, chemical 69–71, 92–4, 101–2, 111–15
 miniaturization of 70
application of chemical knowledge 168
Archer, Frederick Scott 146
Armstrong, Henry 88, 143, 194–5
Arrhenius, Svante 16, 61, 68, 87–8, 125–6, 207
arsenic 122–3, 148
aspirin 14, 155, 179
astrophysics 126–7
Atacama Desert 181
atomic theory 3–7, 10, 30–34, 42, 43–44, 48, 62–4
atomic weights 4–6, 23, 33, 42–4, 56–7, 64–5, 70, 81–2
 agreement on 52–3
Atwood, Mary Anne 36
Avogadro, Amedeo 5, 42, 47, 64
Avogadro's Hypothesis 42, 47, 52–3

Babbage, Charles 209
Baeyer, Adolf 11, 23, 24, 27, 175, 178, 208
Balard, Antoine-Jérôme 132
Ballard, Edward 165
Ballistite 142
Balmer, Johann Jakob 76
Balzac, Honoré de 222, 236–7, 240
Bancroft, Wilder 17
Barrett, Thomas 172
BASF (company) 27–8, 175–6, 178–9, 184–5, 234
Bayer (company) 28, 129, 175–9, 188
Becquerel, Henri 30, 33, 62–3, 145
Beethoven, Ludwig van 24
Bell, John William 172
Bennet, Abraham 114–15
benzene 1, 10, 54, 175
Beretta, Marco 233
Berlin 1, 46, 213
Berlin, University of 19, 102–3, 200, 203, 207, 208, 231–2
Bernard, Claude 123, 132
Bernays, Albert 131
Berthelot, Marcellin 6, 13, 15–16, 36–7, 60, 86–7, 123, 132–3
Berthollet, Claude-Louis 95
Berzelian formulas 11, 68, 76–8
Berzelius, Jacob 3, 7–8, 11–13, 44–7, 64, 65, 70–1, 74, 77, 121, 125
Bessemer, Henry 29
the Bible 238
biochemistry 121, 207
biology's links to chemistry 119, 121, 123
Biot, Jean-Baptiste 55, 60, 111–12
Birkbeck, George 210
Birkeland, Kristian 184
Bjerrum, Niels 76
black smoke 156–7
bleach 11, 76–78
blood stains 122, 230
blowpipe analysis 68–70, 75, 106–7; breathing techniques for 106–7
'blue pots' 104
Bohr, Niels 32, 76
boiling points, measurement of 84, 89, 105–6
Bolton, W.B. 146
Boltwood, Bertram 36
bonding 33–4, 63–4

Bonn, University of 102–3
Bosch, Carl 185
Brande, William 104, 140
Brant, Sebastian 219, 220
bread 152–3
Breslau, University of 19, 98
Brewster, David 209
Britain 3, 6, 16, 20, 23, 24, 25–6, 27, 36–7, 58, 62, 76, 86, 99, 100, 104, 118, 122, 124–5, 130–1, 133–37, 140–8, 150–3, 154–5, 156–7, 162–6, 169–172, 174, 177, 189, 192–5, 196–200, 205–6, 208–9, 209–210, 212, 228, 230
British Association for the Advancement of Science (BAAS) 58, 209–10
Brodie, Benjamin 6, 99–100, 200
Brownian motion 7
Browning, Robert 221–2
Bruegel, Peter the Elder 220
Brunner, John 26, 177
Brunner, Mond & Company 177
Brush, George 203
Buchner, Eduard 121
Buchner, Joseph 14
Buchner, Ludwig 123
Buff, Heinrich 85
Bunsen, Robert 15, 19, 75, 62, 97–9, 103, 107–9, 126, 136, 144, 202, 205–6, 215
Bunsen burner 75–6, 109, 206
Burke, John Butler 34
business practices 188–9
Butlerov, Aleksandr 10, 21–2

Cagliostro 227–8
Cahen, Emile 148
Calcium carbide 184
Cambridge University 20, 76, 143, 196, 200, 207
candles 134, 146
Cannizzaro, Stanislao 5, 52, 57, 64
capital, sources of 189
carbon dioxide 126
carborundum 184
Carlsberg Brewery 189
Caro, Heinrich 27–8, 174–8
Caro, Nikodem 184–5
Casanova, Giacomo 228

cathode rays 30, 62
Caventou, Joseph Bienaimé 13–14
"Caxton" 229
celluloid 127, 187
cellulose 187
Chandler, Charles 21, 214
Chandler, William 214
Chaptal, Jean-Antoine 95
Chaucer, Geoffrey 219
chemical change 42, 60, 91
    neither creating nor destroying matter 41
    using heat 93
chemical engineering 207
Chemical History of a Candle 134, 193
chemical industry 2, 38
    culture of 129–30
    growth and diversification of 24–30
chemical journals 214–15
"Chemical Revolution" 67, 224
Chemical Society of London 110, 144, 210–14
chemicals, need for 169
Chemische Fabrik Griesheim 183
chemistry
    benefits from better knowledge of 140, 166
    branches of 19, 38, 89–90, 93, 206–7, 215
    definition of 91
    as an exact science 91–2
    fundamental basis for modern industry 27
    growth as a science, as a profession and as an industry 2
    increasingly seen as a new science with a central position 117–18, 131, 135–6
    not always regarded as a scholarly discipline 18
    popularization of 130–135
    turning points in the history of 38–39
    unification of the discipline 37, 51–2, 57, 64–5
chemistry kits 107
chemists
    career paths for 38
    numbers of 2, 17–18, 38, 141, 168, 215
    in the public sphere 118, 130–6
    showing a widening range of interests 117
    use of the word 210–11
    what they actually did 67, 89
Chevreul, Michel Eugène 84–5, 119, 128–9
Chile 181
Chincha Islands War (1864–1866) 181
chlorine 8, 177, 183, 184
chloroform 14–15
cholera 162
Christ Church laboratory, Oxford 100
Christian, Robert 122
Ciba (company) 176
cinchona bark 178
civil authorities 140
clinical chemistry 122
coal 155–7, 168, 173–4
coal tar 14, 26–7, 38, 82, 83, 168, 174, 178, 186, 187
cocaine 155
Cochrane, Archibald (Earl Dundonald) 169
coffee 151
collaboration between industry and the academic world 178
Collardeau du Heaume, Charles Félix 105
Collins, Wilkie 222, 228, 235
colour and chemistry 127–8
Columbia University 20–21
combustion analysis 49, 73–74
commercial wealth from chemistry 2
commercialization 140
condoms 148
Comte, Auguste 22, 53
conservation of energy 125
conservation of weight 12
consultants, analytical 140
consumer products 38
Conversations on Chemistry 130, 136, 143, 193
Cooke, Josiah Parsons Jr. 100–1
Cooperative Wholesale Society 172
copper industry 160
cordite 142, 186
counterfeiting 219
Couper, Archibald Scott 53
Courtaulds (company), 187
Crookes, William 58, 75–6, 101, 184, 214
crop yields 120

Cross, Charles F. 187
Crum Brown, Alexander 10–12, 54, 78
crystallography 47, 118
Cumming, James 200
Curie, Marie Skłodowska 30–3, 38, 63, 115, 144–5
Curie, Pierre 30–3, 63, 115, 144–5

Daguerre, Louis 127, 146
Dale, John 174
Dalton, John 2, 4, 7, 33, 42, 43, 47, 62, 64, 124, 174, 210
    disagreements with 33, 43
dangerous commodities 148
Dante Alighieri 219
Darwin, George 34–5
Daubeny, Charles 99–100
David, Jacques-Louis 233
Davis, George 207
Davy, Humphry 4, 14, 41, 45–7, 60, 65, 70, 109, 118, 120, 125, 130, 133, 154, 192–3
    invention of miners' safety lamp 193
Deane, Thomas 99
de Boisbaudran, Paul-Emile Le Coq 58, 60, 76
demonstrations, chemical 118, 133–5
Derby, Lord 158
Descartes, René 224
Devonshire Commission 195
Dewar, James 76, 133, 142, 186, 194
Dickens, Charles 226
Dingley Tariff (1897) 177, 183
Dittmar, William 199
Döbereiner, Johann Wolfgang 5, 57
Dostoyevsky, Fyodor 236
Dow Chemical Company 183
Doyle, Arthur Conan 230
Draper, Henry 127
Draper, John 62, 75, 127
Du Pont (company) 186
Duhem, Pierre 6
Duisberg, Carl 28, 178–9
Dulong, Pierre Louis 52
Dumas, Alexandre 227, 236, 237
Dumas, Jean Baptiste 8, 19–20, 50, 57, 64, 78, 207
Dürer, Albrecht 220
dyestuffs 2, 26–9, 129–30, 173–4, 177–8, 179

dyes and fashion 28
dynamite 142, 186
Dynamit AG (Alfred Nobel & Company) 184, 186

the Earth, age of 35–6
Eastman, George 127, 146
École Normal Supérieure (Paris) 133
École Polytechnique (Paris) 94–5, 96, 202, 205, 207
Ede, Robert Best 135
Edison, Thomas 147
educational system 139–41, 194–205
    colleges and universities in Britain 196–200; in Germany 200–1; in France 201–2; In the United States 202–4; in Japan 204–5
    purpose of 96
    for women 143–4
    laboratory instruction 205–6, 207–8
    secondary schools 194–6
Ehrlich, Paul 14, 24, 148–50, 154, 179
Eichengrün, Arthur 179
Einstein, Albert 7
electricity generation 182–4
electroaffinity 34
electrochemical dualism 3, 8, 45–8, 51, 77–8, 88–9, 125
electrochemistry 45–7, 60, 125
electrolysis in industry 45, 47, 183–4
electrons 32, 33, 62–4
electroscopes 114–15
elements
    classification of 57
    Lavoisier's definition of 69
Eliot, Charles 196
Elkington, James 183
Elkington & Co. 146
Emmet, John 95–6
enantiomorphism 112
energetics 6–7, 61–2
Engel'gardt, A.N. 21
entrepreneurship 168
environmental issues 155–61, 231–2
    in Britain 156–7
    in the United States 157
    In France 161
enzymes 121, 124
equipment, chemical *see* apparatus and instrumentation

"equivalents", chemical 4, 44, 64, 72
Erasmus of Rotterdam 219
Erdmann, Otto 19
Erlenmeyer, Emil 21
Esson, William 86, 88
ether and ethification 81
eudiometers 110
Euripides 228
evening classes 192
everyday life, chemistry in 130–1, 145–53
Everett, Edward 100
exhibitions 141–2
experience in industry 169
explosives 142, 186
Eyde, Samuel 184
Eymericus, Nicolaus 219

Faculté de Médecine (Paris) 206
Fahlberg, Constantin 187
Falkenstein, Paul von 102
Familiar Letters on Chemistry (*Chemische Briefe*) 118, 131, 136
Faraday, Michael 37, 44, 46–7, 60, 64, 65, 103–4, 106–7, 109, 114, 118, 131, 133–5, 140, 193–4, 210
 contributions to chemistry 193–4
Farmer, Thomas 171
Faust legend 218, 220, 222, 226
Ferdinand II, King of the Two Sicilies 171
fermentation 121, 124
fertilizer 120, 179–82
Figuier, Louis 132
"fine" chemicals 169
First World War 24, 166
Fischer, Emil 11, 24, 79, 114, 124, 207, 208
food additives 187–8
foodstuffs 150–3, 179–180
forensic medicine 122
formulas, chemical 3–4, 8, 42, 44, 53–5, 64, 76–81
Fortin, Nicolas 94
Fourcroy, Antoine-François de 95–6, 195
fractional crystallization and fractional distillation 85
France 3–4, 6, 8, 13, 16–17, 19–20, 22, 23, 24, 30, 58, 96, 118–19, 121, 122, 132, 144, 153, 158, 161, 168, 189, 195–6, 201–2, 206, 213

Frank, Adolph 180, 185
Frank, Albert 185
Frankenstein (novel) 125, 218, 221–6, 232, 236
Frankland, Edward 6, 51–2, 64–5, 133, 162, 164, 194, 198, 211
Fraunhofer, Joseph 62, 75, 107
Frasch, Herman 171
Freer, Paul 91–2, 115
Fremy, Edmond 119
French Chemical Society 213
French Revolution 236
Fresenius, Carl Remigius 71–3
Friedel, Charles 23
Friend, John 100
fume hood/cupboard 206
furnaces 104–5

Gage, John 128
Gager, C. Stuart 35
Gahn, Johan Gottlieb 70
Galloway, Robert 107
gas mantles 147
gas street lighting 146, 227
gases, study of 126
Gattermann, Ludwig 84
Gaudin, Marc Antoine Augustine 52
Gay-Lussac, Joseph Louis 4, 18, 42, 49, 70, 96, 110
Geigy (company) 176
Geissler, Heinrich 84, 106
Geneva Conference on chemical nomenclature (1892) 23
Gerhardt, Charles 22, 50–1, 64
German Chemical Society 213
Germany 3–8, 13, 17–19, 27–8, 38, 65, 72, 96–100, 122, 129, 133, 136, 144, 157, 168, 174, 176–80, 188, 195–6, 200–1, 213
 primacy of 24, 177–80, 208
Gesner, Abraham 146–7
Gibbs, Antony, & Sons 181
Gibbs, Josiah Willard 61, 65, 207
Gibbs, Oliver Wolcott 203
Gibbs, William 181
Giessen, University of 18–9, 20, 71, 84, 85, 96, 98, 99, 100, 133, 151, 156, 197, 198, 202, 203, 205–6, 207, 214
Gilbert, John Henry 179

Gilman, Daniel Coit 21
glass blowing 70, 95, 105
glass production 172–3
glassware 70
Goethe, Johann Wolfgang von 24, 127–8, 136, 217, 220, 222
gold, value of 36
Godwin, William 221
Gossage, William 25, 158, 170
Göttingen, University of 18, 19, 97, 144, 155, 202, 203, 205, 208
Graebe, Carl 83, 175
Graham, Thomas 197, 210
gravimetry 49, 72–5
Great Exhibition (1851) 141–2
Griess, Peter 27
Griffin, John Joseph 131
Grignard, Victor 202
Grimm Brothers 223
"grouping" reagents 71
guano 180–1
Guldberg, Cato Maximillian 86–8
gunpowder 186
Gutenberg, Johannes 220
Gutzkow, Karl Ferdinand 235
Guyton de Morveau, Louis Bernard 205

Haber, Fritz 24, 185
Hall, Charles Martin 30, 183
Hall, Julia 30
Halm, Friedrich 222
Hammett, Louis Plack 89
Hancock, Thomas 148
Hantzsch, Arthur 11, 79
Harcourt, Augustus George Vernon 86, 88
Harcourt, William Vernon 209
Harvard University 17, 20, 100–1, 196, 203, 206, 239
Hassall, Arthur Hill 150–2
Hatakeyama, Yoshinari 204–5
Haüy, René Just 47
Hawthorne, Nathaniel 222
hazards in the workplace 25, 165–6
Heidelberg, University of 15, 19, 21, 62, 75, 84, 98–9, 109, 144, 198–9, 202, 203, 206, 208
"heavy" chemicals 169
Helm, Georg 6–7
Helmholtz, Hermann von 123, 125

Henry, William 91–2
Hermetic Order of the Golden Dawn 36
heroin 14, 174
Héroult, Paul Louis-Toussaint 30, 183
Herschel, John 75
Hess, Germain 15, 114
Hill, Octavia 164
Hinrichs, Gustav 5
Hitchcock, Ethan Allen 36
Hodgkin, Thomas 214
Hoechst (company), 28, 175, 178, 179, 188
Hoffmann, Felix 14
Hoffmann, Friedrich 179
Hofman, Paul 97
Hofmann, August Wilhelm von 1, 12, 20, 26, 78, 83–4, 101–2, 110, 144, 174, 175, 197–8, 206
  the Hofmann elimination 83
Holmes, Arthur 36
Hopkins, Frederick Gowland 207
Horsford, Eben 20, 100, 203, 206
household commodities 145–8
Huggins, William 62, 126–7
Humboldt, Wilhelm von 200
Huygens, Christiaan 111
Hyatt, John Wesley 187
hydroelectricity 189
hypothetico-deductive method 79–82, 89

Imperial College, London 198
Imperial College of Engineering (Tokyo) 205
India 27, 130
indigo 27, 130, 178
Industrial Revolution 167
industry, focus of work on the needs of 209
instrument makers 69–70, 105
internationalism 22–4
ionic theory 16, 61, 88
ionists 16, 61
isomers and isomerism 7–10, 47, 50, 55, 77, 79
isomorphism 46–7, 50, 118
isotopes 33, 38, 63

Jacobson, Jacob Christian 189
Japan 168, 182, 185, 204–5

Jefferson, Thomas 95–6
Johns Hopkins University 21, 187, 196, 203–4, 214
Johnson, Samuel 203
Jonson, Ben 219
journals (see chemical journals)
Jewett, Frank 30
Jones, Francis 149
Joy, Charles 20–1
Jung, C. G. 218
Japp, Frances 124

Kahlenberg, Louis 17
Kaliapparat apparatus 19, 49, 74–5, 78, 89, 97, 110–11, 115, 207
Kalle & Company 178
Kant, Immanuel 24
Karlsruhe conference (1860) 5, 23, 52–3, 57, 64
Kassel Polytechnical School 97
Kazan, University of 21–2
Kekulé, August 1–2, 5, 9–10, 52–7, 64, 78, 102, 175, 202, 208
Kelvin, Lord 36
Kepler, Johannes 4
kerosene 147
King, Victor 11
King Edward VI High School for Girls, Birmingham 143
King's College, London 197
Kirchhoff, Gustav 15, 62, 75–6, 107–9, 126, 136
Knox, James 170
Koch, Robert 154
Kohlrausch, Friedrich Wilhelm Georg 60
Kolbe, Hermann 13–14, 37–8, 55, 102
Kopp, Hermann 15, 36–7, 60, 84–5, 132, 207

laboratories 91–104
  academic or *artisanal* 70
  equipped with running water 85
  design and evolution of 93–103
  for industrial research 28, 129–30
  practical instruction in 205–6, 215
  setting-up of 103–4
  techniques employed in 89, 93
  tradition in 67
Ladenburg, Albert 84

Lafarge, Marie 122–3
Lamy, Claude Auguste 76
The Lancet 151
Landolt, Hans 15
Laplace, Pierre Simon de 114
laudanum 150
Laurent, Auguste 9, 22–3, 50–1, 64, 118
Lavoisier, Antoine-Laurent 3, 23, 41, 46, 49, 69, 93, 103, 114, 120, 125, 132, 233, 238
  execution of (1794) 94, 233
Lavoisier, Marie-Anne Paulze 233
Lavoisierian revolution 131
Lawes, John Bennett 179–80
Lawrence, Abbott 206
Le Bel, Joseph Achille 55, 60, 112
Leblanc, Nicholas 24–6, 158
Leblanc process 140, 157–8, 161, 165, 169–72, 177, 184
Le Chatelier, Henri Louis 62
lectures 118, 133
Lee, Matthew 100
legal disputes, scientific evidence in 140
Leipzig 180, 213
Leipzig, University of 16, 17, 19, 102, 103, 126
Lermontova, Yulia 144
Le Rossignol, Robert 185
Lever Brothers 172
Lewis, Gilbert N. 34, 63–4
Liebermann, Carl 83, 175
Liebig, Justus von 8, 12, 18–20, 22, 47, 49–51, 64, 71, 74–5, 77–8, 83, 92, 96–7, 103, 110, 118, 119–21, 131–2, 136, 151, 156, 170, 179–80, 197, 202, 203, 205, 207, 214, 235
Lister, Joseph 121–2, 154
"Lit and Phils" 192
lighting, artificial 26, 102–3, 146–7, 173, 184, 227
lithium 109
Little, Arthur 207
Liveing, George 76, 200
Lockyer, Norman 62, 76, 127
Losh, William 169–70
Löwig, Carl 19
Lübeck conference (1895) 6–7
Ludwig, Carl 123

Ludwigshafen 175, 234
Lyon, University of 202

Macintosh, Charles 148, 170
MacNeven, William 96
"mad alchemist" image 218–20, 228, 237
"mad scientist" image 217–8, 221, 225, 226, 229, 231, 235, 238
Magendie, François 123
male domination of chemistry 38, 143–4
Malus, Etienne-Louis 111
Manchester 156–7, 164, 168
Manchester University 144, 198
manometers 114
Mansfield, Charles 26, 101
Marburg, University of 19, 97, 102, 198
Marcet, Jane 130–1, 135, 136–7, 143, 193
Marignac, Jean Charles Galisand 52
markets, control of 189
Marlowe, Christopher 220
Marsh, James 122, 148
Maskelyne, Nevil Story 99
mass production 39
matches 148, 165–6
materialism 123–4
mathematics and mathematization 16, 88, 117
Mathieson Alkali Works 183
mauve 26, 82, 129, 174, 197
Maximum Work, principle of 87
mechanics' institutes 192
medicines 13–14, 121–2, 154–5
Mégnié, Pierre Bernard 114
melting points, measurement of 84, 89, 106
Mendeleev, Dmitri 5–6, 42, 57–60, 65, 68, 82
Mercer, John 128
Mersey Chemical Works 172, 177
Meyer, Ernst von 36–7
Meyer, Julius Lothar 5, 52, 57
Meyer, Stefan 38
Meyer, Victor 79
military training for student chemists 95
Miller, William Allen 75
Mitscherlich, Eilhard 46, 118
Mittasch, Alwin 185
Mohr, Friedrich 72
Moleschott, Jacob 123
Mond, Ludwig 26, 177, 194

Monsanto (company) 187–8
moral perversion 226–8
Morgan, Thomas H. 35
morphine 13
Morton, William 14, 154
Moseley, Henry 32–3
Muir, M.M. Patterson 37
Mulder, Gerardus 119
Müller, Franz 106
Muller, Hermann J. 35
Munich, University of 202, 203, 208
Murray, Sir James 179
Muspratt, James 25, 170–1
Muspratt, Sheridan 83–4, 151, 206

nation-states 22
national heroes 22
national styles 22
nationalism 22, 24
natural history 67–8, 89
natural philosophy 46, 117–18, 134
Nernst, Walther 124
Newcastle Chemical Company (Allhusen) 170, 177
Newlands, John 5
Newton, Sir Isaac 4, 127–8, 224
Nicol prisms 112
Niemann, Albert 155
Niépce, Joseph 127
nitrogen fixation 182, 184–5
Nobel, Alfred 142, 145, 186
Nobel prizes 33, 61, 142, 145, 186, 202
noble (inert) gases 6, 58–60, 126, 194
nomenclature of chemistry 23, 41, 95, 131
Norsk Hydro (company) 184
notation, chemical 8–11, 46, 54, 61, 68, 76–81
Noyes, Arthur 17

Odling, William 200
oil lamps 146
Oresme, Nicole 219
Orfila, Mateu 122–3
organic chemistry 7–15, 37–8, 48–52, 53–7, 206–7
    rise of 7–15
    structural theory of 1–2, 9–10, 27, 53–4

and agriculture 120, 179–182
and clinical chemistry 122
Ostwald, Wilhelm 6–7, 16–17, 23, 24, 38, 61–2, 64, 86, 87, 93, 103, 114, 125–6, 133, 207
Owens College, Manchester 20, 198–9, 208, 212
Oxford University 20, 37, 58, 86, 99–100, 196, 200, 208–9

Paley, William 124
Palmerston, Lord 156
paper tools 77–82, 89
Paracelsus 223
Parkes, Alexander 187
Parkes, Samuel 135
Paschen, Friedrich 76
Pasteur, Louis 11, 55, 60, 112, 118–9, 121, 124, 154
patents 28, 30, 129, 140, 142, 145, 148, 158, 176, 178, 180, 186, 187, 188, 214
Paul, Jean 222
Pavesi, Angelo 52
Pears (company) 172
Peel, Sir Robert 197
Pelletier, Pierre-Joseph 13–14, 119
periodic table 5–7, 33, 38, 42, 57–60, 63, 68, 82
Perkin, William Henry 26, 82, 101, 129, 173–4, 176, 189, 198, 233
Perkin, William Henry Junior 200, 208–9
Petit, Alexis Thérèse 52
Petrarch, Francesco 219
Pfaff, Christoph Heinrich 70–1
pharmaceutical industry 55, 117, 178–9
Pharmaceutical Society of Great Britain 211
Pharmacists, training 18
philosopher's stone 132
photography 127–8, 146
physical chemistry 16–17, 60–2, 65, 85–9, 93, 125, 207
    cultural divide from those trained in traditional methods 88
physics 2, 6–7, 33
    links to chemistry 33, 76, 88, 117–18, 125, 136
Pilkington (company) 172
Pilkington, William 172

Pittsburgh Plate Glass Company 173
Planck, Max 32
plants, physiological effects of 13
Plattner, Karl 107
Playfair, Lyon 128, 198
"playing God" 222, 236–7, 238
pluralism in chemical practice 89–90
Poe, Edgar Allan 222
Poincaré, Henri 6
poisoning 122, 148
polarimetry and polarimeter 111–15
pollutants 161–4
Pope, William 200
popularization of chemistry 131–3, 136
Port Sunlight 172
portraiture 233–4
positivism 64, 123
precision 44
process theology 124
professionalization 2, 17–22, 140, 191, 194, 209–15
protein 119
Prout, William 4, 37, 43 119, 124–5
Prout's Hypothesis 4, 37, 43–4
psychology 225–6
public health 18, 139

quantitative chemistry 41, 93
quantum theory 32, 76
Queeny, John F. 187–8
quinine 14

Raabe, Wilhelm 231–2
radical theory 8, 49–51, 64, 78
radioactivity 3, 30–3, 36–8, 42, 62–3
    broader impact of 34–7
radium 31, 33, 34
Ramsay, William 6, 16, 32, 37, 58–60, 63, 76, 126, 127, 194
Raoult, François 15
Raspe, Rudolf Erich 228
rayon 187
Reeve, Arthur Benjamin 231
reductionism 121
Regnault, Victor 85
regulation of chemicals 140
Reich, Ferdinand 76
religious beliefs 124, 237–8
Remsen, Ira 21, 196, 203–4

Rensselaer Polytechnical Institute 95, 100, 203, 206
research schools 201, 208, 207–208
Richards, Theodore 17
Richet, Charles 123
Richter, H. Theodor 76, 107
Richter, Jeremiah 69
rivalry between countries 22–4
Roberts, Dale & Company 174
Röntgen, Wilhelm 30, 62
Roosevelt, Theodore 188
Roscoe, Henry 20, 109, 164, 198–9, 208, 212
Rose, Heinrich 71
Royal College of Chemistry (RCC) 20, 101, 162, 174, 179, 197, 198, 208
Royal Institute of Chemistry 211
Royal School of Mines 198
Royal Institution (RI), London 41, 104, 118, 130, 133, 135, 143–4, 154, 192
Royal Society 209
Ruskin, John 164
Russia 21, 144
Rutgers College 204
Rutherford, Ernest 31–2, 33, 62–3
Rydberg, Johannes Robert 76

Sabatier, Paul 202
saccharin 187–8
Saint-Claire Deville, Henri 133–4
St. Helens Plate Glass Company 172
St. Petersburg, University of 21–2, 144
St. Rollox Works 171
Sakurai, Joji 81
Salvarsan 14, 148, 149, 150, 154, 179
Samuelson Report (1868) 194–5
"sausage" formulas 54, 78
Savene, Henri 148
Sayce, B.J. 146
Scherer, Johann Josef 122
schools, chemistry in 194–6
Schorlemmer, Carl 199
science
   chemistry's status as 19, 166, 215
   type of education offered 96, 101–2
Scott, Reginald 219
Senderens, Jean-Baptiste 202
Sertürner, Friedrich Wilhelm 13
Serviss, Garrett P. 36

Seurat, Georges 128–9
sewage 162
Shakespeare, William 228
Shelley, Mary 125, 218, 221–5, 237
Sherlock Holmes stories 218, 230–1
Signac, Paul 128
Silliman, Benjamin 107, 203
silver-plating 146
silver fulminate 47
Simpson, James 14–15, 154
slide rule of chemical equivalents 44
Smedley, Ida 143
Smith, Robert Angus 156–7, 160–1
smoke abatement 156–7, 161
Snow, John 14, 162
soap 172
Society of Chemical Industry (SCI) 211
Society of German Naturalists and Physicians 210, 213
societal impact of chemistry 140–2, 218
societies for chemistry 211–13, 215
   local 141
soda (sodium carbonate) 24–5
sodium nitrate (Chilean soda, Chilean nitrate) 181, 132
Soddy, Frederick 31–3, 37, 63
Sokolov, N.N. 21
Solvay, Ernest 25–6, 177, 189
Solvay Process Company 177
solvents 184
Somerville, Mary 143
Sorbonne 132, 207
spectroscope and spectroscopy 15, 62, 75–6, 107–11, 126
Spies, Johann 220
Spitzweg, Carl 233
spontaneous generation 118, 124
steel 29–30
Steinheil, Carl August 108–9
stereochemistry 11, 42, 55–6, 79, 93
Stevenson, Robert Louis 226
Stieler, Robert Friedrich 234
stoichiometry 4, 7, 15, 62, 69, 77, 93, 109–10
stopcock burettes 72
Stromeyer, Friedrich 18, 97, 205
Strutt, John William 6, 194
Strutt, Robert 36, 58
students working in laboratories 95–6

Sturgeon, Theodore 223
Substitution 50–1, 55, 78, 83, 118
suicide 150
Swan, Joseph 147
synthesis 12–13, 37–38, 68, 75
  "constructive" synthesis 83
synthetic products 190
syphilis 148–50, 179
Syx, Max 36

tea 151
Talbot, Henry Fox 75, 127, 146
teaching methods for chemistry 95, 101, 197
teaching-research laboratories 19–22, 38, 49, 95, 101, 196–206
Teniers, David the Younger 220
Tennant, Charles 25, 170
Tennant, Charles, & Company 170
Tennant, John 171
tetrahedral carbon atom 11, 55, 79, 112
textbooks of chemistry 70–1, 131
Tharsis Sulphur & Copper Company 171
Thayer, Sylvanus 95
Thenard, Louis-Jacques 49, 70, 96, 133
theory-driven chemistry 169
Theosophical Society 36
thermochemistry 60, 114
thermodynamics 16, 117, 125
  chemical 42, 60–1, 65
thermometers 84–5
Thirlmere (Lake) 164
Thomas, Sydney Gilchrist 29
Thompson, Benjamin 192
Thomsen, Julius 15–16, 60, 86
Thomson, Joseph J. 32, 33, 62
Thomson, Thomas 3, 43–4, 205, 207, 210
Thoreau, Henry David 231
Thorpe, Edward 199
titrimetry 72
TNT 186
Tokyo University 204
toothpaste 150
Torricelli, Evangelista 114
Toulouse, University of 202
toxicology 122–3
training of chemists 2, 17–22, 196–206
transmutation, radioactive 31, 33, 35–7
transmutation, alchemical 132

Travers, Morris 76
'triads' 5
Trommsdorff, Johann 18
Turgenev, Ivan 236
Turner, Edward 197, 205
Tyndall, John 126, 133, 194
type theory 9, 51, 53, 64

Union Carbide Company 184
Union College (NY) 20
United Alkali Company 177, 183
United States 17, 20–1, 22, 100–101, 107, 120, 130, 143, 144, 147, 153, 164, 166, 168, 171, 172, 177, 180, 183, 186, 187, 188, 195, 196, 200, 202, 204, 205, 206, 207, 212–3
  Department of Agriculture 153, 180
  Military Academy (USMA) 95
  Physical chemistry in 17
unity of chemistry 37–38, 49, 51, 52, 57, 65
universities 2, 196–205
  benefit from increased state funding 19–20, 115
University College, London (UCL) 20, 197, 204
urbanization 139, 164, 166
urea, synthesis of 13, 37–38, 123

vacuum pumps 85
valence 9–10, 51, 53, 61, 63–4
van 't Hoff, Jacobus Henricus 11–12, 16, 55–6, 60, 61–2, 68, 112–13, 125–6, 142, 207
Vauquelin, Louis-Nicholas 119
venereal diseases 148, 154
Verguin, François 27, 174
Verne, Jules 227, 229
Victoria, Queen 14–15, 26
Virginia, University of 95–6
visual representation of chemistry 232–4
vitalism 12–13, 123–4
Vogt, Carl 123
Volta, Alessandro 3, 114, 125
Voltaic pile 41
Vyrnwy valley 164

Waage, Peter 86–8
Wakley, Thomas 151

Walker, John 148
Walker Alkali Works 169
Walton, Frederick 145
Wanklyn, James Alfred 162, 194
warfare 140, 142
    chemical 24, 142, 166
Washburn, Frank Sherman 185
"waste" products 38
water supply 162
    in Britain 162–164
    in the United States 164–5
water vapour 126
Waterhouse, Alfred 198
Webster, John 100
Weiditz, Hans the Younger 220
Weisbach, Carl von 147
Weishaupt, Adam 236
Wells, H.G. 36, 222, 223, 228–9
Wells, Horace 154
Weltzien, Karl 52
Werner, Abraham Gottlob 47
Werner, Alfred 11, 55–6, 79
Whewell, William 46–7
Widnes 159, 161, 170–1, 172, 212
Wieland, Christoph Martin 222
Wiesbaden Agricultural Institute 72
Wiley, Harvey 153, 188
Wilhelmy, Ludwig Ferdinand 86

Williamson, Alexander 6, 23, 53, 81–2, 204
Willson, Thomas Leopold 184
Willstätter, Richard 24
Wislicenus, Johannes 7, 11, 79, 208
Witt, Otto N. 27, 176
Wöhler, Friedrich 8, 12–13, 19, 37–8, 47, 50, 77, 83, 123, 202, 208
Wollaston, William Hyde 4, 44–5, 75, 107
women
    as chemists 38, 142–5
    in public life 130–131, 135, 139–40, 166
    in higher education 143–4
wood cellulose 186–7
Woodward, Benjamin 99
working conditions 25, 165–6
Wright, Joseph 233
Wright, William Valentine 172
Wurtz, Adolphe 5, 6, 20, 23, 24, 52, 57, 132, 206

x-rays 30, 34, 62

Yale University 61, 107, 203
Young, James 146

Zamminer, Friedrich 85
Zinin, Nikolai 21
Zürich, University of 19